Elements of Environmental Chemistry

Elements of Environmental Chemistry

Molly Miller

STATES
ACADEMIC PRESS
www.statesacademicpress.com

Published by States Academic Press
109 South 5th Street,
Brooklyn, NY 11249, USA
www.statesacademicpress.com

Elements of Environmental Chemistry
Molly Miller

International Standard Book Number: 978-1-63989-174-0 (Hardback)

Cataloging-in-Publication Data

Elements of environmental chemistry / Molly Miller.
p. cm.
Includes bibliographical references and index.
ISBN 978-1-63989-174-0
1. Environmental chemistry. 2. Chemistry. I. Miller, Molly.
TD193 .E44 2021
628.5--dc23

Contents

Preface

The branch of science which focuses on all the chemical and biochemical events which occur in natural places is known as environmental chemistry. It attempts to study the reactions, sources, effects and transport of the chemical species which are present in the soil, water and air. It also deals with the effects of human activities and biological activities on these systems. Environmental chemistry is a multidisciplinary field which encompasses the principles of soil chemistry, atmospheric chemistry and aquatic chemistry. It finds applications in the detection of pollutants such as heavy metals, aromatic carbons and organometallic compounds in water bodies and organic systems. Phenomena such as algal bloom and eutrophication are also studied under it. The topics included in this book on environmental chemistry are of utmost significance and bound to provide incredible insights to readers. Its objective is to give a general view of the different areas of environmental chemistry, and its applications. Those in search of information to further their knowledge will be greatly assisted by this book.

A foreword of all chapters of the book is provided below:

Chapter 1 - The layer of gases surrounding the Earth is known as the atmosphere. The study of chemical processes which take place in water, air and living environments, as well as the impacts of human activities upon them are studied under the discipline of environmental chemistry. This chapter will provide a brief introduction to the different aspects of Earth's atmosphere and chemistry such as atmospheric structure and Earth's energy budget; **Chapter 2 -** The geosphere of the Earth encompasses the rocks and minerals on Earth as well as the abiotic portions of soil and the fossilized skeletons of animals. The chemical characteristics of soil are also studied within this discipline. This chapter has been carefully written to provide an easy understanding of the varied facets of geosphere and its chemistry such as internal structure of Earth and soil formation; **Chapter 3 -** The total amount of water on a planet is termed as the hydrosphere. It includes the water present on the surface, in the air and under the ground. A few areas of focus within this field are water cycle and water pollution. The topics elaborated in this chapter will help in gaining a better perspective about these aspects of hydrosphere and its chemistry; **Chapter 4 -** The part of the environment which has been constructed or changed by the humans for their habitat and use comprise the anthroposphere. A few areas of study within this field are anthropogenic activities, industrial ecology and green chemistry. This chapter closely examines these key concepts of anthroposphere to provide an extensive understanding of the subject; **Chapter 5 -** The mixture of solid particles and gases within the air is called air pollution. Some of the major types of air pollutants are sulphur oxides, nitrogen oxides, carbon monoxide, carbon dioxide and volatile organic compounds. This chapter discusses in detail these air pollutants as well as greenhouse effect and climate change; **Chapter 6 -** The contamination of water bodies through various reasons is known as water pollution. The branch of study which focuses on the impact of manufactured chemicals and other materials either manmade or natural on aquatic animals is termed as aquatic toxicology. The topics elaborated in this chapter will help in gaining a better perspective about water pollution and toxicology as well as the different sources of water pollution.

At the end, I would like to thank all the people associated with this book devoting their precious time and providing their valuable contributions to this book. I would also like to express my gratitude to my fellow colleagues who encouraged me throughout the process.

Molly Miller

1

Earth's Atmosphere and Chemistry

The layer of gases surrounding the Earth is known as the atmosphere. The study of chemical processes which take place in water, air and living environments, as well as the impacts of human activities upon them are studied under the discipline of environmental chemistry. This chapter will provide a brief introduction to the different aspects of Earth's atmosphere and chemistry such as atmospheric structure and Earth's energy budget.

Atmospheric Structure

The gaseous area surrounding the planet is divided into several concentric strata or layers. About 99% of the total atmospheric mass is concentrated in the first 20 miles (32 km) above Earth's surface. Historical outline on the discovery of atmospheric structure.

Thermal Structure

Atmospheric layers are characterized by variations in temperature resulting primarily from the absorption of solar radiation; visible light at the surface, near ultraviolet radiation in the middle atmosphere, and far ultraviolet radiation in the upper atmosphere.

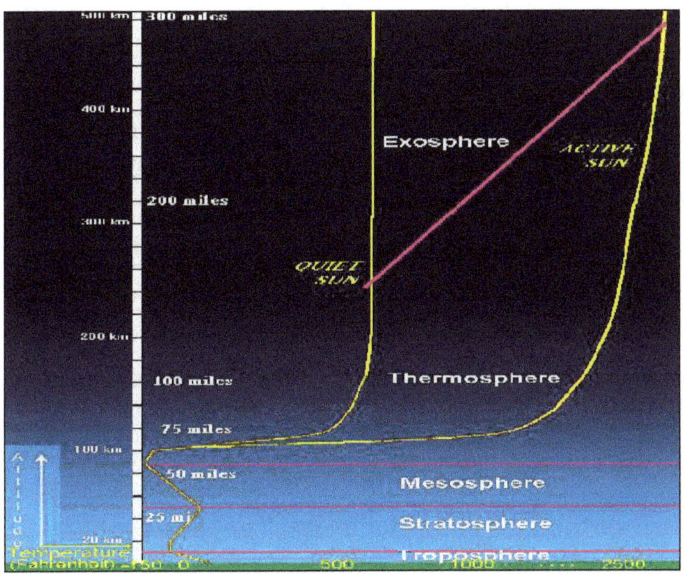

Troposphere

The troposphere is the atmospheric layer closest to the planet and contains the largest percentage (around 80%) of the mass of the total atmosphere. Temperature and water vapor content in the troposphere decrease rapidly with altitude. Water vapor plays a major role in regulating air temperature because it absorbs solar energy and thermal radiation from the planet's surface. The troposphere contains 99 % of the water vapor in the atmosphere. Water vapor concentrations vary with latitude. They are greatest above the tropics, where they may be as high as 3 %, and decrease toward the Polar Regions.

All weather phenomena occur within the troposphere, although turbulence may extend into the lower portion of the stratosphere. Troposphere means "region of mixing" and is so named because of vigorous convective air currents within the layer.

The upper boundary of the layer, known as the tropopause, ranges in height from 5 miles (8 km) near the poles up to 11 miles (18 km) above the equator. Its height also varies with the seasons; highest in the summer and lowest in the winter.

Stratosphere

The stratosphere is the second major strata of air in the atmosphere. It extends above the tropopause to an altitude of about 30 miles (50 km) above the planet's surface. The air temperature in the stratosphere remains relatively constant up to an altitude of 15 miles (25 km). Then it increases gradually to up to the stratopause. Because the air temperature in the stratosphere increases with altitude, it does not cause convection and has a stabilizing effect on atmospheric conditions in the region. Ozone plays the major role in regulating the thermal regime of the stratosphere, as water vapor content within the layer is very low. Temperature increases with ozone concentration. Solar energy is converted to kinetic energy when ozone molecules absorb ultraviolet radiation, resulting in heating of the stratosphere.

The ozone layer is centered at an altitude between 10-15 miles (15-25 km). Approximately 90 % of the ozone in the atmosphere resides in the stratosphere. Ozone concentration in this region is about 10 parts per million by volume (ppmv) as compared to approximately 0.04 ppmv in the troposphere. Ozone absorbs the bulk of solar ultraviolet radiation in wavelengths from 290 nm - 320 nm (UV-B radiation). These wavelengths are harmful to life because they can be absorbed by the nucleic acid in cells. Increased penetration of ultraviolet radiation to the planet's surface would damage plant life and have harmful environmental consequences. Appreciably large amounts of solar ultraviolet radiation would result in a host of biological effects, such as a dramatic increase in cancers.

Mesosphere

The mesosphere a layer extending from approximately 30 to 50 miles (50 to 85 km) above the surface is characterized by decreasing temperatures. The coldest temperatures in Earth's atmosphere occur at the top of this layer, the mesopause, especially in the summer near the pole. The mesosphere has sometimes jocularly been referred to as the "ignorosphere" because it had been probably the least studied of the atmospheric layers. The stratosphere and mesosphere together are sometimes referred to as the middle atmosphere.

Thermosphere

The thermosphere is located above the mesosphere. The temperature in the thermosphere generally increases with altitude reaching 600 to 3000 F (600-2000 K) depending on solar activity.

This increase in temperature is due to the absorption of intense solar radiation by the limited amount of remaining molecular oxygen. At this extreme altitude gas molecules are widely separated. Above 60 miles (100 km) from Earth's surface the chemical composition of air becomes strongly dependent on altitude and the atmosphere becomes enriched with lighter gases (atomic oxygen, helium and hydrogen). Also at 60 miles (100 km) altitude, Earth's atmosphere becomes too thin to support aircraft and vehicles need to travel at orbital velocities to stay aloft. This demarcation between aeronautics and astronautics is known as the Karman Line. Above about 100 miles (160 km) altitude the major atmospheric component becomes atomic oxygen. At very high altitudes, the residual gases begin to stratify according to molecular mass, because of gravitational separation.

Exosphere

The exosphere is the most distant atmospheric region from Earth's surface. In the exosphere, an upward travelling molecule can escape to space (if it is moving fast enough) or be pulled back to Earth by gravity (if it isn't) with little probability of colliding with another molecule. The altitude of its lower boundary, known as the thermopause or exobase, ranges from about 150 to 300 miles (250-500 km) depending on solar activity. The upper boundary can be defined theoretically by the altitude (about 120,000 miles, half the distance to the Moon) at which the influence of solar radiation pressure on atomic hydrogen velocities exceeds that of the Earth's gravitational pull. The exosphere observable from space as the geocorona is seen to extend to at least 60,000 miles from the surface of the Earth. The exosphere is a transitional zone between Earth's atmosphere and interplanetary space.

Magneto-Electronic Structure

The upper atmosphere is also divided into regions based on the behavior and number of free electrons and other charged particles.

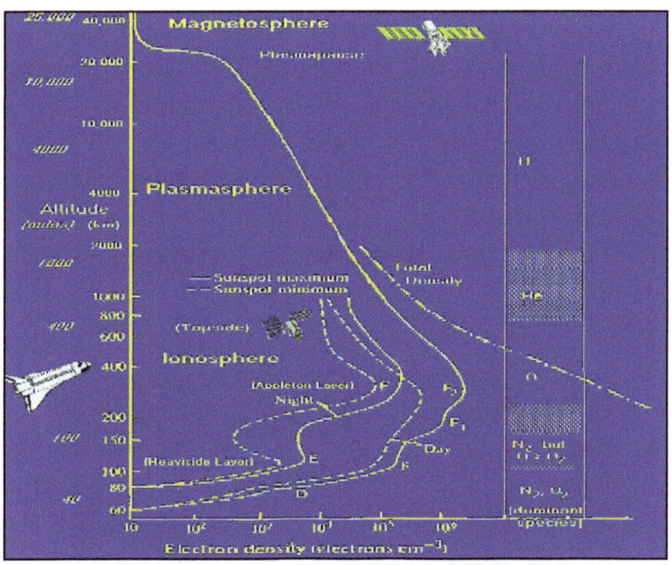

Ionosphere

The ionosphere is defined by atmospheric effects on radio wave propagation as a result of the presence and variation in concentration of free electrons in the atmosphere.

- D-region is about 35 to 55 miles (60 - 90 km) in altitude but disappears at night.
- E-region is about 55 to 90 miles (90 - 140 km) in altitude.
- F-region is above 90 miles (140 km) in attitude. During the day it has two regions known as the F1-region from about 90 to 115 miles (140 to 180 km) altitude and the F2-region in which the concentration of electrons peaks in the altitude range of 150 to 300 miles (around 250 to 500 km). The ionosphere above the peak electron concentration is usually referred to as the Topside Ionosphere.

Plasmasphere

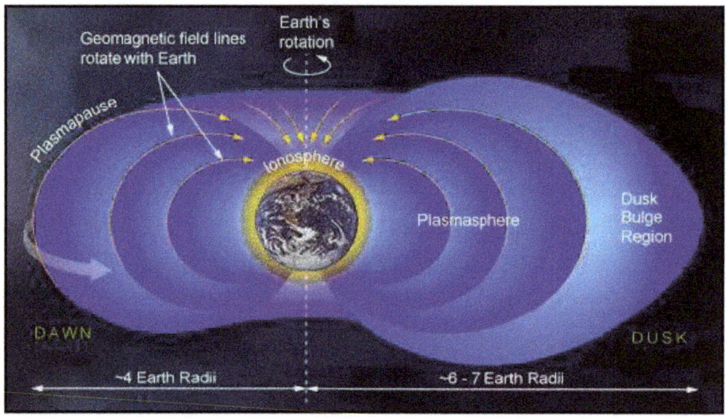

The plasmasphere is not really spherical but a doughnut-shaped region (a torus) with the hole aligned with Earth's magnetic axis. [In this case the use of the suffix -sphere is more in the figurative sense of a "sphere of influence".] The Earth's plasmasphere is made of just that, plasma, the fourth state of matter. This plasma is composed mostly of hydrogen ions (protons) and electrons. It has a very sharp edge called the plasmapause. The outer edge of this doughnut over the equator is usually some 4 to 6 Earth radii from the center of the Earth or 12,000-20,000 miles (19,000-32,000 km) above the surface. The plasmasphere is essentially an extension of the ionosphere. Inside of the plasmapause, geomagnetic field lines rotate with the Earth. The inner edge of the plasmasphere is taken as the altitude at which protons replace oxygen as the dominant species in the ionospheric plasma which usually occurs at about 600 miles (1000 km) altitude. The plasmasphere can also be considered to be a structure within the magnetosphere.

Magnetosphere

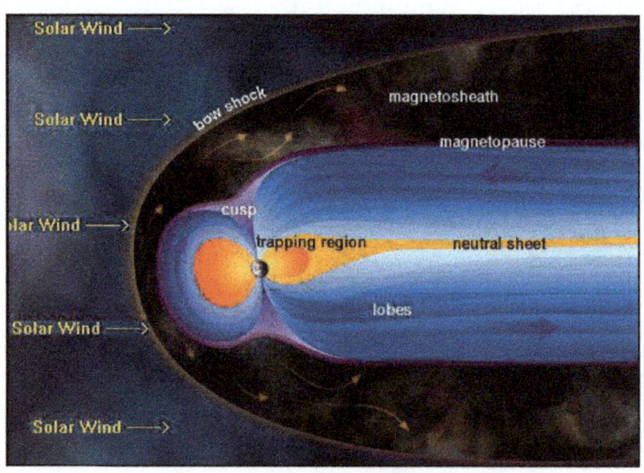

Ozone Formation and Destruction

Stratospheric ozone (O_3) is produced by the combination of an oxygen atom (O) with an oxygen molecule (O_2). The basic steps to formation are given in image below:

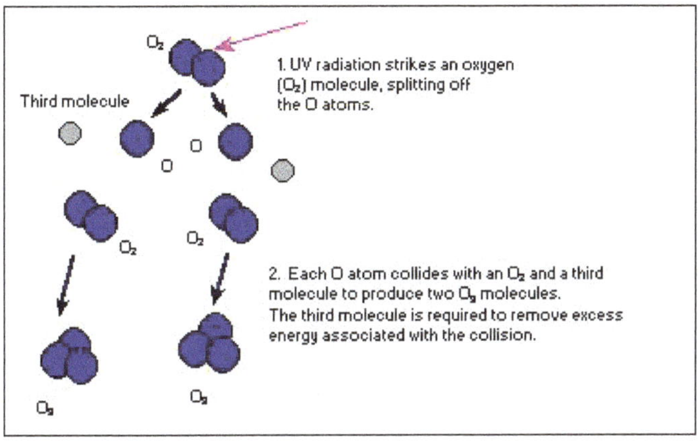

In the above diagram, oxygen atoms are represented as dark blue circles. This reaction is written in chemical equations as:

O_2 + UV => O + O

2O + 2 O_2 + third molecule => 2O_3 + third molecule

Net Reaction: 3O_2 + UV => 2O_3

UV radiation is also involved in the destruction of O_3.

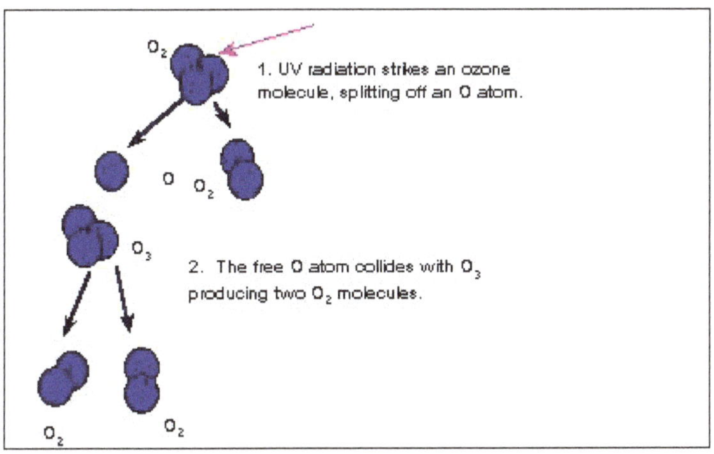

This destruction is expressed as:

O_3 + UV => O + O_2

O + O_3 => 2 O_2

Net Reaction: 2O_3 + UV => 3O_2

UV radiation plays a crucial role in the formation and destruction of ozone. UV amounts are greatest in the tropical regions, thus it is not surprising that most of the destruction and production of ozone occurs in the tropical stratosphere.

In 1970 Dr. P. Crutzen proposed the following catalytic reaction that results in the destruction of O_3:

$$X + O_3 => XO + O_2$$

$$O_3 + UV => 2O_2$$

$$O + XO => X + O_2$$

Net Reaction: $2O_3 + UV => 3O_2$

In this sequence of reactions, X is an atom or molecule that acts as a catalyst to convert O_3 to O_2. Note that X does not change in the net reaction and so it can continue to destroy O_3 molecules. The important radicals represented by X include chlorine (Cl), hydroxyl (OH), nitric oxide (NO), and bromine (Br). There is a delicate balance between the production and destruction of O_3, resulting in what is referred to as an O_3 shield that protects us from high energy UV radiation. This natural balance has recently been disrupted by human activities.

One molecule that can serve as the catalyst molecule X is chlorine (Cl). But how does Cl get into the stratosphere? In the 1930s useful chemical compounds known as chlorofluorocarbons (CFCs) were produced for use in refrigeration, air conditioning, solvents, aerosol spray cans, and Styrofoam™ puffing agents. They are very stable in the troposphere with lifetimes of approximately 100 years. This long lifetime allows CFCs that are emitted near the surface to be carried by the winds upward. In the stratosphere CFCs are dissociated by UV light producing chlorine atoms. The destruction of O_3 then follows with the following chemical reactions:

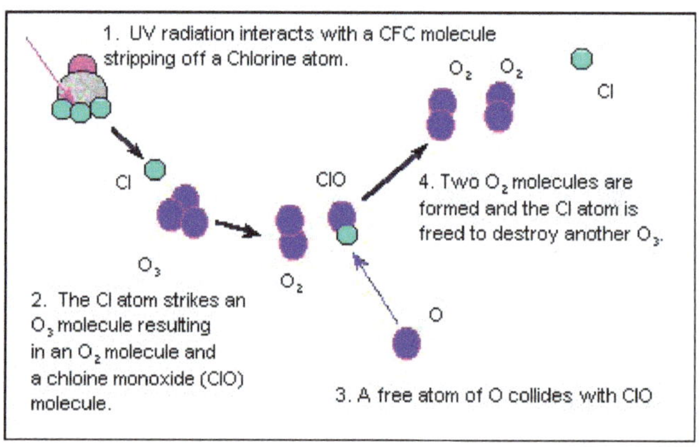

Fortunately chlorine atoms (represented as light blue circles) do not remain in the stratosphere for centuries. If the use of CFCs and other ozone-destroying chemicals are banned, it is hoped that ozone depletion may be reduced. The term ozone depletion means that the destruction of O_3 exceeds the creation of O_3.

When present together in the stratosphere, chlorine (Cl) and ozone quickly react to produce chlorine oxide. Bromine can also act as a catalyst to destroy stratospheric ozone.

In 1995 Drs. Paul Crutzen, Mario J. Molina and F. Sherwood Rowland won the Nobel Prize in chemistry for their work concerning the formation and decomposition of ozone.

Ozone Depletion

CFC molecules are made up of chlorine, fluorine and carbon atoms and are extremely stable. This extreme stability allows CFC's to slowly make their way into the stratosphere (most molecules decompose before they can cross into the stratosphere from the troposphere). This prolonged life in the atmosphere allows them to reach great altitudes where photons are more energetic. When the CFC's come into contact with these high energy photons, their individual components are freed from the whole. The following reaction displays how Cl atoms have an ozone destroying cycle:

$Cl+O_3 \rightarrow ClO+O_2$ (step 1)

$ClO+O. \rightarrow Cl+O_2$ (step 2)

$O_3+O. \rightarrow 2O_2$ (Overall reaction)

Chlorine is able to destroy so much of the ozone because it acts as a catalyst. Chlorine initiates the breakdown of ozone and combines with freed oxygen to create two oxygen molecules. After each reaction, chlorine begins the destructive cycle again with another ozone molecule. One chlorine atom can thereby destroy thousands of ozone molecules. Because ozone molecules are being broken down they are unable to absorb any ultraviolet light so we experience more intense UV radiation at the earth's surface.

From 1985 to 1988, researchers studying atmospheric properties over the South Pole continually noticed significantly reduced concentrations of ozone directly over the continent of Antarctica. For three years it was assumed that the ozone data was incorrect and was due to some type of instrument malfunction. In 1988, researchers finally realized their error and concluded that an enormous hole in the ozone layer had indeed developed over Antarctica. Examination of NASA satellite data later showed that the hole had begun to develop in the mid 1970's.

The ozone hole over Antarctica is formed by a slew of unique atmospheric conditions over the continent that combines to create an ideal environment for ozone destruction. Because Antarctica is surrounded by water, winds over the continent blow in a unique clockwise direction creating a so called "polar vortex" that effectively contains a single static air mass over the continent. As a result, air over Antarctica does not mix with air in the rest of the earth's atmosphere.

Antarctica has the coldest winter temperatures on earth, often reaching -110 F. These chilling temperatures result in the formation of polar stratospheric clouds (PSC's) which are a conglomeration of frozen H_2O and HNO_3. Due to their extremely cold temperatures, PSC's form an electrostatic attraction with CFC molecules as well as other halogenated compounds. As spring comes to Antarctica, the PSC's melt in the stratosphere and release all of the halogenated compounds that were previously absorbed to the cloud. In the Antarctic summer, high energy photons are able to photolyze the halogenated compounds, freeing halogen radicals that then catalytically destroy O_3. Because Antarctica is constantly surrounded by a polar vortex, radical halogens are not able to be diluted over the entire globe. The ozone hole develops as result of this process.

Resent research suggests that the strength of the polar vortex from any given year is directly correlated to the size of the ozone hole. In years with a strong polar vortex, the ozone hole is seen to expand in diameter, whereas in years with a weaker polar vortex, the ozone hole is noted to shrink.

Ozone Depleting Substances

The following substances are listed as ozone depleting substances:

Substance	Ozone- depletion potential
chlorofluorocarbon-11 (CFC–11)	1.0
chlorofluorocarbon-12 (CFC–12)	1.0
chlorofluorocarbon-13 (CFC–13)	1.0
chlorofluorocarbon-111 (CFC–111)	1.0
chlorofluorocarbon-112 (CFC–112)	1.0
chlorofluorocarbon-113 (CFC–113)	0.8
chlorofluorocarbon-114 (CFC–114)	1.0
chlorofluorocarbon-115 (CFC–115)	0.6
chlorofluorocarbon-211 (CFC–211)	1.0
chlorofluorocarbon-212 (CFC–212)	1.0
chlorofluorocarbon-213 (CFC–213)	1.0
chlorofluorocarbon-214 (CFC–214)	1.0
chlorofluorocarbon-215 (CFC–215)	1.0
chlorofluorocarbon-216 (CFC–216)	1.0
chlorofluorocarbon-217 (CFC–217)	1.0
halon-1211	3.0
halon-1301	10.0
halon-2402	6.0
carbon tetrachloride	1.1
methyl chloroform	0.1
hydrochlorofluorocarbon-22 (HCFC–22)	0.05
hydrochlorofluorocarbon-123 (HCFC–123)	0.02
hydrochlorofluorocarbon-124 (HCFC–124)	0.02
hydrochlorofluorocarbon-141(b) (HCFC–141(b))	0.1
hydrochlorofluorocarbon-142(b) (HCFC–142(b))	0.06

Kármán Line

Space

It's the final frontier and also happens to be where NOAA operates the Nation's environmental and meteorological satellites.

But where is "space" exactly? This may seem like a simple question, but any answer beyond "up" may be more complicated than you think. Although most people are generally in agreement that space begins when Earth's atmosphere ends— where exactly that is depends on who you ask.

International law states that outer space shall be free for exploration and use by all, but there is no definitive law stating where national air space actually ends and outer space begins. This leaves the door open for a variety of interpretations.

A common definition of space is known as the Kármán Line, an imaginary boundary 100 kilometers (62 miles) above mean sea level. In theory, once this 100 km line is crossed, the atmosphere becomes too thin to provide enough lift for conventional aircraft to maintain flight. At this altitude, a conventional plane would need to reach orbital velocity or risk falling back to Earth.

The world governing body for aeronautic and astronautic records, the Fédération Aéronautique Internationale (FAI), and many other organizations use the Kármán Line as a way of determining when space flight has been achieved.

The U.S. military and NASA define space differently. According to them, space starts 12 miles below the Kármán Line, at 50 miles above Earth's surface. Pilots, mission specialists and civilians who cross this boundary are officially deemed astronauts.

If we define getting to space by the strictest of terms—that is, by escaping Earth's atmosphere completely—you may have to travel over 600 miles or more to the outermost layer of the atmosphere to reach it. Here, the atmosphere becomes incredibly thin and starts to give way to the stronger, more violent solar winds of the sun.

This way of defining space complicates things a bit, though. At that altitude the International Space Station (orbiting between 205 to 270 miles up), the space shuttle (which orbited 200 miles up) and some of NOAA's polar-orbiting satellites (orbiting 540 miles up) would not be considered spacecraft.

The concept of "Freedom of Space" began in the 1950's with the dawn of the space age, the Cold War and President Dwight D. Eisenhower. With the threat of nuclear war looming, President Eisenhower and his advisers sought to ensure international acceptance of "freedom of space" for the unspoken use of reconnaissance satellites.

His interpretation of space treated it much like the high seas. Outside of national territorial airspace, nations were free to conduct peaceful space operations without concern for international borders. Well aware of the President's unstated motivation to operate high altitude spy satellites safely, the Soviet Union adamantly opposed the idea, seeking to establish much higher airspace boundaries.

Fig: William H. Pickering, James van Allen, and Wernher von Braun hold aloft a full-scale model of Explorer I, the United States' first satellite, at a press conference announcing its successful entry into orbit.

Despite such protests, Eisenhower announced that the United States would launch a scientific Earth satellite as part of the 1957 International Geophysical Year (IGY).

The U.S. hoped that a science satellite, already part of the IGY plan and far less controversial than a spy satellite, would make the world's first foray into space and set the precedent for freedom of space.

Ironically, it was the Soviet Union who inadvertently set the precedent with the launch of Sputnik-1, the world's first artificial satellite. As Sputnik, and the satellites that followed during the IGY, crossed international borders without protest, the concept of Freedom of Space" was tentatively established.

Basic Properties of Atmosphere

The structure of the atmosphere and the properties of its different atmospheric layers greatly determine the atmospheric chemistry. Our atmosphere is unique and its properties are such that among all the known planets and other heavenly bodies, life exists and sustains only on Earth. The major constituents are nitrogen, oxygen and CO_2 followed by other gases in smaller amounts and traces. About 99.9% of mass air is concentrated in troposphere and stratosphere (up to ~ 50 km) and half of the mass of air is in first 6 km from the Earth's surface.

The importance of oxygen needs no elaboration. All living organism, including humans, require oxygen for respiration, and the conversion of chemical energy stored in food for use by living beings for life sustaining functions. Carbon dioxide is in photosynthesis to produce food and replenish the oxygen supply. Nitrogen is important in several ways, such synthesis of proteins, nucleotides, fertilizers, etc.

Mass of the Atmosphere

Atmospheric pressure is defined as the force per unit area exerted on a unit area of the surface by the weight of the air above the surface of the Earth. The mean atmospheric pressure at sea level is 1 atmosphere or 760 torr or 1.013 × 105 Pa or Pascal or 1.013 bar. At the surface of the Earth, mean global atmospheric pressure P_{earth} is 984 hPa, slightly less than at mean sea level pressure due to higher elevation of Earth.

The mass of the overlying atmosphere can be calculated as follows:

P_{earth} is defined as by Equation below:

$$P_{earth} = \text{Total atmospheric force / Total surface area},$$

$$P_{earth} = mag / 4\pi R\rho^2$$

Where g = acceleration due to gravity = 9.8 ms^{-2}, R_{earth} = radius of the Earth = 6400 km and P_o = 0.984 x 105 Pa = 0.984 x 10^5 kg m^{-1} s^{-2}. The Eq. on rearranging becomes:

$$ma = 4P_{earth}\pi\ R\rho^2 / g$$

On substituting these values in above Eq., we obtain mass of the atmosphere, m_a = 5.2 x 1018 kg.

Chemical Composition

The relative average concentrations of major gaseous constituents of air at sea level are in table below:

Gases	Percent/ppm
Nitrogen	78.084%
Oxygen	20.946%
Argon	0.9345%
Carbon dioxide	0.039%
Others	Traces

The trace atmospheric constituents are responsible for air pollution and each of the gaseous constituents, such as, CO, SO_2, NO_x, methane, VOCs etc.

Vertical Profile of Atmospheric Pressure/Variation in Atmospheric Pressure

With increase in altitude, the pressure continuously decreases due to decrease in gravitational attraction. Te variation of pressure with altitude is shown in Fig. below.

At a height of 5 km from Earth surface, pressure is nearly half the value at surface. This thinning of air with altitude induces breathlessness. At Mt. Everest air pressure drops to about 265 mbar. So mountaineers need to use bottled oxygen. That is why the air pressure in aircrafts is maintained artificially at 1atmosphere. Some features of atmosphere are as follows:

- The gaseous components are compressible.
- The gas molecules do not settle down under the influence of gravity because translational kinetic energy competes with sedimentation forces and keeps the gas molecules afloat. Had

it not been tall the gases would have been virtually within few centimeters at the surface with vacuum overhead.

- The pressure and density decrease exponentially with increase in altitude as defined by Barometric Law equation.

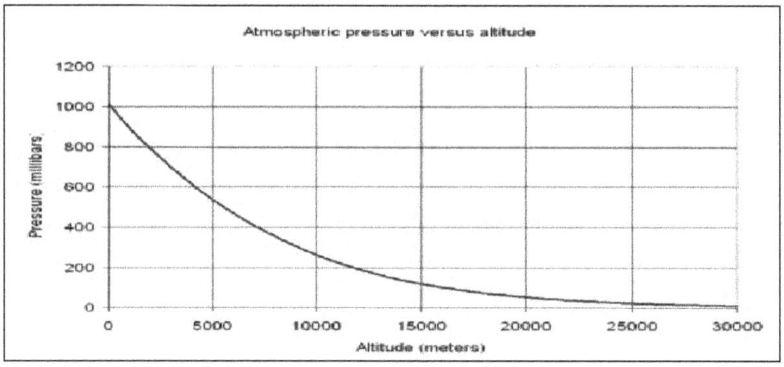

Fig: The variation of atmospheric pressure with altitude.

$$P_z = P_{earth}\ e^{-MgZ/RT}$$

Where,

P_{earth} = Pressure at Earth surface

P_z = Pressure at height z

m = Molecular mass of the gas

T = temperature in Kelvin

The term scale height, H, is very much in use and it is defined as:

$$H = RT/mg$$

On substituting the value of H in Equation above, we obtain Equation:

$$P_z = P_{earth}\ e^{-Z/H}$$

The logarithmic form of log of the above Equation is:

$$\ln P_z = \ln P_{earth} - Z/H$$

On rearranging $P_z = P_{earth}\ e^{-Z/H}$ become $\ln P_z = \ln P_{earth} - Z/H$

$$Z = H \ln \left(P_{earth}/P_z\right)$$

Since the pressure at Earth surface is equal to one atmosphere, equation $\ln P_z = \ln P_{earth} - Z/H$ becomes $Z = H \ln \left(P_{earth}/P_z\right)$

$$Z = -H \ln P_z$$

It can be seen from Eq. $\ln P_z = \ln P_{earth} - Z/H$ that a scale height is a distance over which the pressure decreases by a factor of e, i. e., ~2.71828. Assuming the temperature to be invariant with height (which is not true), the plot of Z and ln Pz is expected to be linear having a Slope = - H. The value of H is found to be ~7.4. This value corresponds to the scale height at mean atmospheric temperature of 250 °C. We have assumed the temperature to be constant, but in reality it is not. However, the effect of change in T is small because it is used in Kelvin scale. The real ln P versus altitude plot shows come curvature due to variation in temp. So H varies from 6 km at 210 K to 8.5 km at 290 K.

The scale height has been calculated from Eq. using s 28.97 (~ 29) as the mean molar mass of dry air, m.

Air Mixing Ways: Homosphere and Heterosphere

Interestingly up to an altitude of ~100 km, air behaves as a single homogeneous component of molecular mass, M = 28.8 (~29) and the composition of gases remains unchanged. This is contrary to Barometric law, which requires the concentration of heavier species such as N_2, O_2, to decrease more rapidly than lighter elements like He and H_2 with increase in height. Above 100 km, the concentration of heavier elements starts decreasing more rapidly than the lighter elements and air composition starts changing. And the percentage of lighter gases starts increasing.

This observed behavior in air composition above and below 100 km is dependent on which of the two mixing ways, viz., mixing by molecular diffusion and macroscopic mixing dominate.

Mixing by Molecular Diffusion

The main features are:

- It discriminates the molecules on the basis of their molecular masses in accordance with Barometric law.
- Gravitational attraction tries to separate the molecules on molecular scale during diffusion.
- Mean free path, λ_m, increases on increasing altitude due to decrease in pressure.

$$\lambda_m = \left(1/2^{1/2}\pi d^2\right)\left(kT/P\right)$$

Where,

d = diameter,

k = Boltzmann constant,

P = pressure.

Mean free path is the average distance traveled by a moving particle between successive collisions.

Macroscopic Mixing/Bulk Fluid Motion/Turbulent Mixing

This kind of mixing occurs on macro scale due to convection, turbulence, small eddies or fluctuations. It does not discriminate among gas molecules on the basis of their molecular masses. The fluid mixing length is a distance that a fluid parcel will maintain its original character before mixing in the surrounding fluid. Fluid Mixing Length, λ_F, tends to decrease with increase in altitude. Thus it is the distance travelled during a Transport Event. In atmosphere, whereas the molecular diffusion attempts to separate the molecules on the basis of their molecular masses, on the other hand the fluid motion tries to redistribute the gas molecules that are being separated by gravitational attraction.

Up to 100 km, the fluid motions dominate since $\lambda_F > \lambda_m$. And the composition of air remains unchanged. So air behaves as a single component. This part is called Homosphere. Between 100 -120 km, λ_F and λ_m are almost of the same size (0.1 – 1.0 m). This region is known as turbopause.

Above 120 km, mixing by molecular diffusion dominates as λ_m is much larger than λ_F. Here, composition is based on molecular mass. So above 120 km, the proportion of lighter gases starts increasing. This region is called Heterosphere.

At the top of the atmosphere only lighter components such as H, H_2 and He are present. Some of these escape in outer space depending up on escape velocity.

Adiabatic Lapse Rate

The decrease in temperature with increase in height can be calculated as follows. Imagine a rising parcel of dry air. The parcel is in equilibrium with its surroundings, but thermally isolated from it and behaves adiabatically. Since on increasing altitude, the pressure decreases, the parcel of air expands and does work against surroundings consuming its internal energy, E.

We know that:

$$dE = dq + dw$$

And $H = E + PV$

On differentiating,

From above equations:

$$dH = dq + dw + PdV + VdP$$

But conditions are adiabatic so:

$$dq = 0 \text{ and } dw = -PdV$$

Hence,

$$dH = 0 - PdV + PdV + VdP$$

$$dH = VdP$$

We know that:

$$dP = \text{force} / \text{area}$$

$$dP = -\rho gAdz / A = -\rho gdz$$

Where ρ = density, g = acceleration due to gravity, A = area, dz = change in height, force =mass ×g, mass = density× volume and volume = dz × A.

Thus from equation $dP = \text{force} / \text{area}$:

$$dH = -V\rho\, g\, dz$$

The molar heat capacity, Cp, is related to change in enthalpy with temperature as follows.

$$Cp = \left(dH / dT\right)_P \text{ Or } dH = CpdT$$

Heat capacity for molar mass is Cp, and for unit mass (1 g) heat capacity id cp. For unit mass, ρ = 1/ V. Keeping this in view,

$cpdt = -dz\, g$ (Remember the difference between Cp and cp),

Or $\Gamma d = -dT / dz = g / cp$

$\Gamma d = -dT / dz = g / cp$ is known as dry adiabatic lapse rate for a dry parcel. From the values of g and cp(for air), lapse rate is found to be 9.8° C/km. Notably Γ_d does not depend on atmospheric conditions.

Saturated Adiabatic Lapse Rate, T_s

When condensable vapors are present, latent heat is released and this adds complications. For a saturated vapor, decrease in temperature is accompanied by release of latent heat. And so the decrease in temperature with attitude is lower than in the case of dry air parcel. Thus, saturated adiabatic lapse rate, T_s, is lower than T_d. Its value lies in Earth's atmosphere between 4 -7 °C/km.

By comparing the atmospheric lapse rate, Γ_{atm},, with Γ_d, the vertical stability of the atmosphere with respect to buoyancy can be predicted.

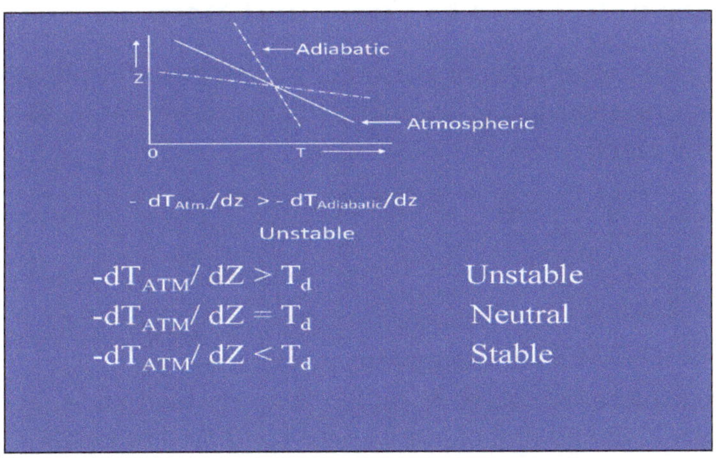

The resistance of the atmosphere to vertical motion is the stability. The degree of stability or instability of the atmosphere at any particular instant determines the upward or downward vertical movement. This depends up on the actual lapse rate is greater or less than the dry adiabatic lapse rate. The departure of this from the dry adiabatic lapse rate determines whether the atmosphere is unstable or stable. The Unstable atmosphere encourages the vertical movement of air. When the air is stable, vertical movement of air is hindered.

In sum, if Γ_{atm},, is less than Γ_d and temperature decreases with altitude, the atmosphere is stable, and if if Γ_{atm},, is less than Γ_d but there is temperature inversion that is temperature increases with altitude then the atmosphere is very stable. On the contrary, if if Γ_{atm},, is more than Γ_d the atmosphere is then unstable, and if Γ_{atm},, is much more than Γ_d then the atmosphere is very unstable. Finally, when Γ_{atm},, = Γ_d atmosphere is neutral. In this case the parcel of air will have same the same density and temperature as the surroundings.

Temperature/Thermal Inversion

Temperature inversion occurs when the instead of decreasing with altitude the temperature starts increasing. It is called temperature inversion. In such a situation, air above the ground is warmer than the air below. Earth has the quality of heating faster as well as cooing faster than the surrounding air. On a clear night,when the air near the ground cools rapidly it loses heat and the Earth surface becomes cooler quickly than the air above because its ability to lose heat slowly it retains and remains warmer than the Earth surface. There are other ways of inversion also.

Temperature inversion greatly influences the meteorology as the atmosphere becomes stable It blocks the dispersion of pollutants by restricting vertical mixing. By not allowing the dispersion of pollutants, inversion helps in the development of the significant amounts of smog. Particularly in winter, many Indian cities see the accumulation of pollutants released by automobiles and other human activities. London's Great Smog and Mexico's similar problems are extreme examples of smog being impacted by the presence of an inversion layer. This is a problem all over the world though and cities like Los Angeles, California; Mumbai, India; Santiago, Chile; and Tehran, Iran, frequently experience intense smog when an inversion layer develops over them. Inversion is involved in other weather phenomena such as freezing rain, thunderstorms and tornadoes also.

Earth's Atmosphere in Relation to Venus and Mars

Wayne has presented an interesting comparison of atmospheres of planets and the reason why Earth sustains life. Table presents the atmospheric concentrations of N_2, O_2 and CO_2 and temperature in Venus, Earth and Mars.

Planet	CO_2, ppm	N_2, ppm	O_2, ppm	Surface temperature
Venus	965000	35000	< 0.3	735 K
Earth	400	780840	209460	288 K
Mars	953200	27000	1300	223 K

In Earth's atmosphere, concentrations of N_2 and O_2 are very high and that of CO_2 very low. Temperature of Venus is very high and that of Mars low as compared to Earth. The concentration of O_2 maintained at high concentration in Earth due to the its replenishment by biological processes for example, photosynthesis in plants., despite its removal by reactions with nitrogen,, H_2 and with a large number of other elements and compounds, The CO_2 at lower level is due to its consumption by plants and many other processes. In fact the presence of biosphere is responsible for current status of Earth. In its absence it would be like Venus and Mars.

Earth's Energy Budget

The Sun doesn't heat the Earth evenly. Because the Earth is a sphere, the Sun heats equatorial regions more than Polar Regions. The atmosphere and ocean work non-stop to even out solar heating imbalances through evaporation of surface water, convection, rainfall, winds, and ocean circulation. This coupled atmosphere and ocean circulation is known as Earth's heat engine.

The climate's heat engine must not only redistribute solar heat from the equator toward the poles, but also from the Earth's surface and lower atmosphere back to space. Otherwise, Earth would endlessly heat up. Earth's temperature doesn't infinitely rise because the surface and the atmosphere are simultaneously radiating heat to space. This net flow of energy into and out of the Earth system is Earth's energy budget.

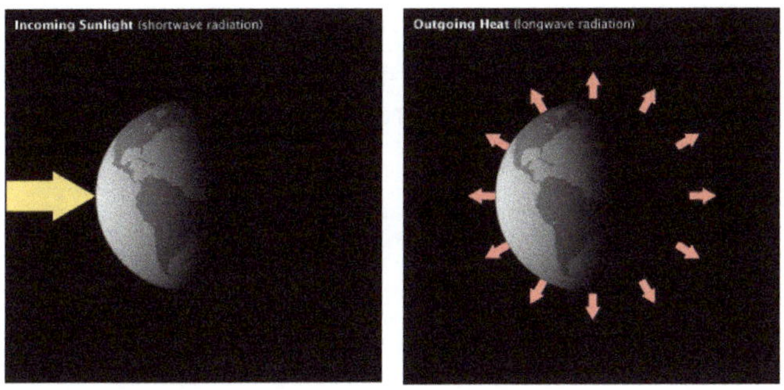

The energy that Earth receives from sunlight is balanced by an equal amount of energy radiating into space. The energy escapes in the form of thermal infrared radiation: like the energy you feel radiating from a heat lamp.

When the flow of incoming solar energy is balanced by an equal flow of heat to space, Earth is in radiative equilibrium, and global temperature is relatively stable. Anything that increases or decreases the amount of incoming or outgoing energy disturbs Earth's radiative equilibrium; global temperatures rise or fall in response.

Incoming Sunlight

All matter in the universe that has a temperature above absolute zero (the temperature at which all atomic or molecular motion stops) radiates energy across a range of wavelengths in the electromagnetic spectrum. The hotter something is, the shorter its peak wavelength of radiated energy is. The hottest objects in the universe radiate mostly gamma rays and x-rays. Cooler objects emit mostly longer-wavelength radiation, including visible light, thermal infrared, radio, and microwaves.

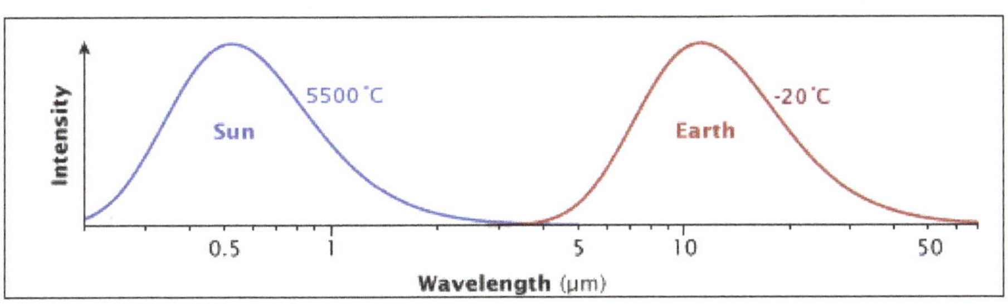

The Sun's surface temperature is 5,500° C, and its peak radiation is in visible wavelengths of light. Earth's effective temperature—the temperature it appears when viewed from space—is -20° C, and it radiates energy that peaks in thermal infrared wavelengths.

The surface of the Sun has a temperature of about 5,800 Kelvin (about 5,500 degrees Celsius, or about 10,000 degrees Fahrenheit). At that temperature, most of the energy the Sun radiates is visible and near-infrared light. At Earth's average distance from the Sun (about 150 million kilometers), the average intensity of solar energy reaching the top of the atmosphere directly facing the Sun is about 1,360 watts per square meter. This amount of power is known as the total solar irradiance. (Before scientists discovered that it varies by a small amount during the sunspot cycle, total solar irradiance was sometimes called "the solar constant").

A watt is measurement of power, or the amount of energy that something generates or uses over time. How much power is 1,360 watts? An incandescent light bulb uses anywhere from 40 to 100 watts. A microwave uses about 1000 watts. If for just one hour, you could capture and re-use all the solar energy arriving over a single square meter at the top of the atmosphere directly facing the Sun—an area no wider than an adult's outstretched arm span—you would have enough to run a refrigerator all day.

Incandescent light bulbs radiate 40 to 100 watts. The Sun delivers 1,360 watts per square meter. An astronaut facing the Sun has a surface area of about 0.85 square meters, so he or she receives energy equivalent to 19 60-watt light bulbs.

The total solar irradiance is the maximum possible power that the Sun can deliver to a planet at Earth's average distance from the Sun; basic geometry limits the actual solar energy intercepted by Earth. Only half the Earth is ever lit by the Sun at one time, which halves the total solar irradiance.

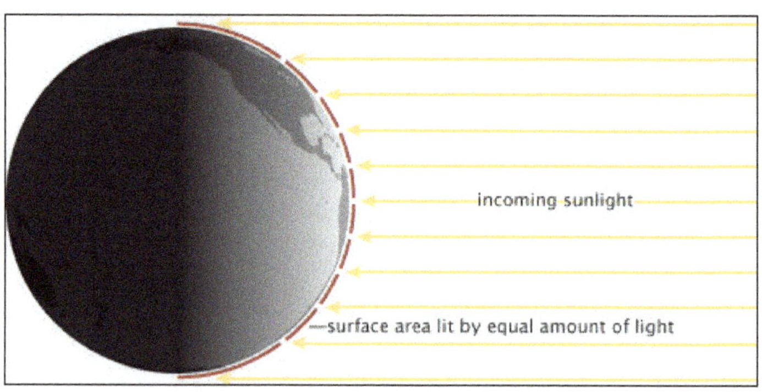

Energy from sunlight is not spread evenly over Earth. One hemisphere is always dark, receiving no solar radiation at all. On the daylight side, only the point directly under the Sun receives full-intensity solar radiation. From the equator to the poles, the Sun' rays meet Earth at smaller and smaller angles, and the light gets spread over larger and larger surface areas.

In addition, the total solar irradiance is the maximum power the Sun can deliver to a surface that is perpendicular to the path of incoming light. Because the Earth is a sphere, only areas near the equator at midday come close to being perpendicular to the path of incoming light. Everywhere else, the light comes in at an angle. The progressive decrease in the angle of solar illumination with increasing latitude reduces the average solar irradiance by an additional one-half.

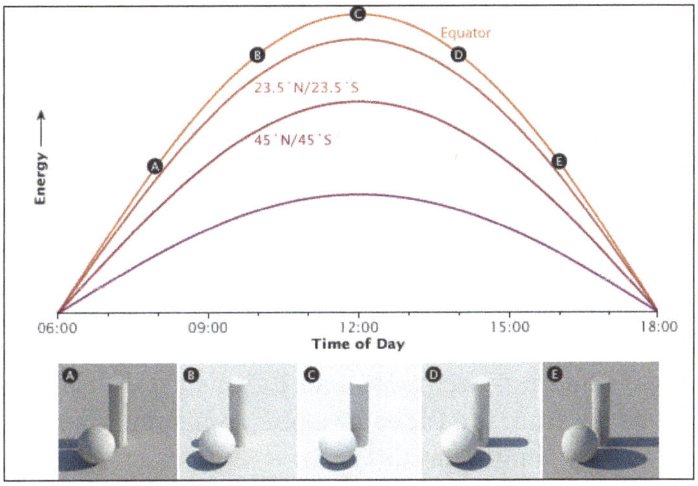

The solar radiation received at Earth's surface varies by time and latitude. This graph illustrates the relationship between latitude, time, and solar energy during the equinoxes. The illustrations show how the time of day (A-E) affects the angle of incoming sunlight (revealed by the length of the shadow) and the light's intensity. On the equinoxes, the Sun rises at 6:00 a.m. everywhere. The strength of sunlight increases from sunrise until noon, when the Sun is directly overhead along the equator (casting no shadow). After noon, the strength of sunlight decreases until the Sun sets at 6:00 p.m. The tropics (from 0 to 23.5° latitude) receive about 90% of the energy compared to the equator, the mid-latitudes (45°) roughly 70%, and the Arctic and Antarctic Circles about 40%.

Averaged over the entire planet, the amount of sunlight arriving at the top of Earth's atmosphere is only one-fourth of the total solar irradiance, or approximately 340 watts per square meter.

When the flow of incoming solar energy is balanced by an equal flow of heat to space, Earth is in radiative equilibrium, and global temperature is relatively stable. Anything that increases or decreases the amount of incoming or outgoing energy disturbs Earth's radiative equilibrium; global temperatures must rise or fall in response.

Heating Imbalances

Three hundred forty watts per square meter of incoming solar power is a global average; solar illumination varies in space and time. The annual amount of incoming solar energy varies considerably from tropical latitudes to polar latitudes. At middle and high latitudes, it also varies considerably from season to season.

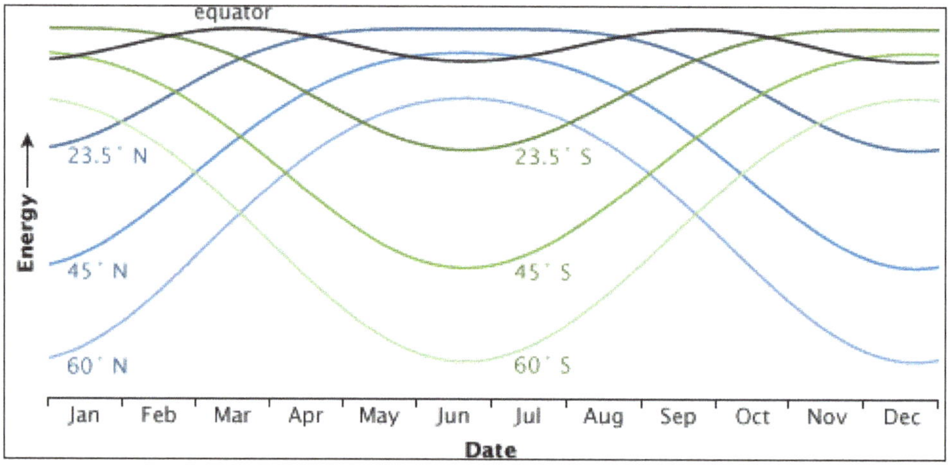

The peak energy received at different latitudes changes throughout the year. This graph shows how the solar energy received at local noon each day of the year changes with latitude. At the equator (gray line), the peak energy changes very little throughout the year. At high northern (blue lines) and southern (green) latitudes, the seasonal change is extreme.

If the Earth's axis of rotation were vertical with respect to the path of its orbit around the Sun, the size of the heating imbalance between equator and the poles would be the same year round, and the seasons we experience would not occur. Instead Earth's axis is tilted off vertical by about 23 degrees. As the Earth orbits the Sun, the tilt causes one hemisphere and then the other to receive more direct sunlight and to have longer days.

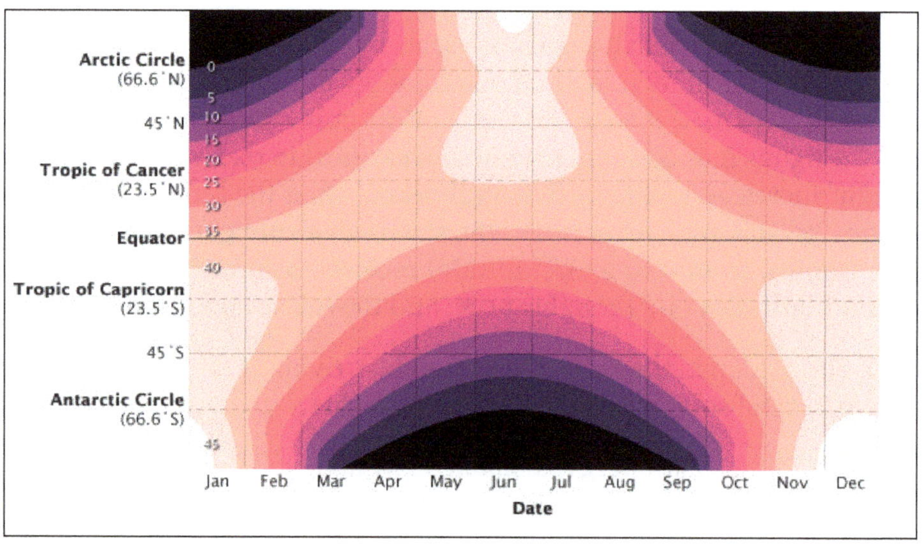

The total energy received each day at the top of the atmosphere depends on latitude. The highest daily amounts of incoming energy (pale pink) occur at high latitudes in summer, when days are long, rather than at the equator. In winter, some polar latitudes receive no light at all (black). The Southern Hemisphere receives more energy during December (southern summer) than the Northern Hemisphere does in June (northern summer) because Earth's orbit is not a perfect circle and Earth is slightly closer to the Sun during that part of its orbit. Total energy received ranges from 0 (during polar winter) to about 50 (during polar summer) mega joules per square meter per day.

In the "summer hemisphere," the combination of more direct sunlight and longer days means the pole can receive more incoming sunlight than the tropics, but in the winter hemisphere, it gets none. Even though illumination increases at the poles in the summer, bright white snow and sea ice reflect a significant portion of the incoming light, reducing the potential solar heating.

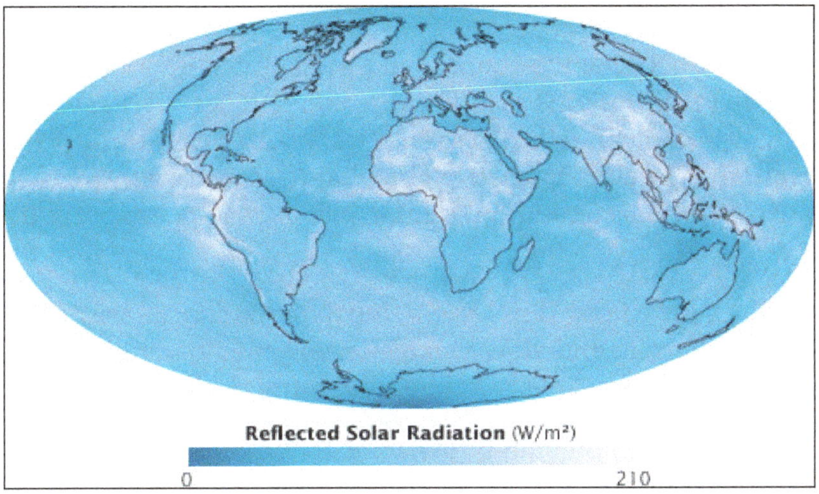

The amount of sunlight the Earth absorbs depends on the reflectivness of the atmosphere and the ground surface. This satellite map shows the amount of solar radiation (watts per square meter) reflected during September 2008. Along the equator, clouds reflected a large proportion of sunlight, while the pale sands of the Sahara caused the high reflectivness in North Africa. Neither pole is receiving much incoming sunlight at this time of year, so they reflect little energy even though both are ice-covered.

The differences in reflectivness (albedo) and solar illumination at different latitudes lead to net heating imbalances throughout the Earth system. At any place on Earth, the net heating is the difference between the amount of incoming sunlight and the amount heat radiated by the Earth back to space. In the tropics there is a net energy surplus because the amount of sunlight absorbed is larger than the amount of heat radiated. In the Polar Regions, however, there is an annual energy deficit because the amount of heat radiated to space is larger than the amount of absorbed sunlight.

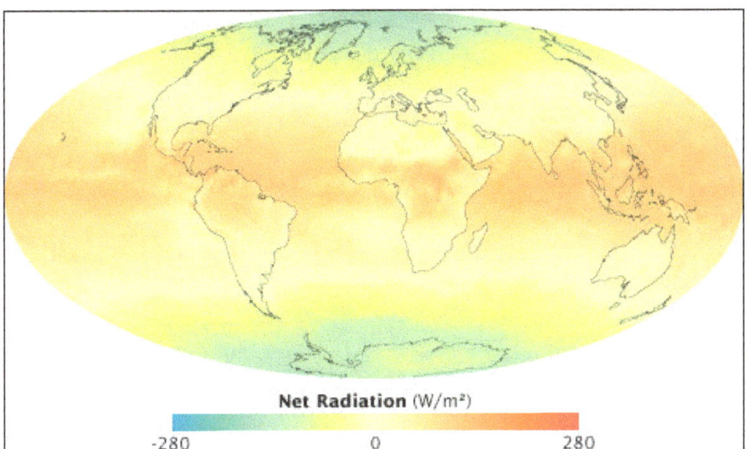

This map of net radiation (incoming sunlight minus reflected light and outgoing heat) shows global energy imbalances in September 2008, the month of an equinox. Areas around the equator absorbed about 200 watts per square meter more on average (orange and red) than they reflected or radiated. Areas near the poles reflected and/or radiated about 200 more watts per square meter (green and blue) than they absorbed. Mid-latitudes were roughly in balance.

The net heating imbalance between the equator and poles drives an atmospheric and oceanic circulation that climate scientists describe as a "heat engine." (In our everyday experience, we associate the word engine with automobiles, but to a scientist, an engine is any device or system that converts energy into motion.) The climate is an engine that uses heat energy to keep the atmosphere and ocean moving. Evaporation, convection, rainfall, winds, and ocean currents are all part of the Earth's heat engine.

Determining exact values for energy flows in the Earth system is an area of ongoing climate research. Different estimates exist, and all estimates have some uncertainty. Estimates come from satellite observations, ground-based observations, and numerical weather models.

Earth's heat engine does more than simply move heat from one part of the surface to another; it also moves heat from the Earth's surface and lower atmosphere back to space. This flow of incoming and outgoing energy is Earth's energy budget. For Earth's temperature to be stable over long periods of time, incoming energy and outgoing energy have to be equal. In other words, the energy budget at the top of the atmosphere must balance. This state of balance is called radiative equilibrium.

About 29 percent of the solar energy that arrives at the top of the atmosphere is reflected back to space by clouds, atmospheric particles, or bright ground surfaces like sea ice and snow. This energy plays no role in Earth's climate system. About 23 percent of incoming solar energy is absorbed in the atmosphere by water vapor, dust, and ozone, and 48 percent passes through the atmosphere and is absorbed by the surface. Thus, about 71 percent of the total incoming solar energy is absorbed by the Earth system.

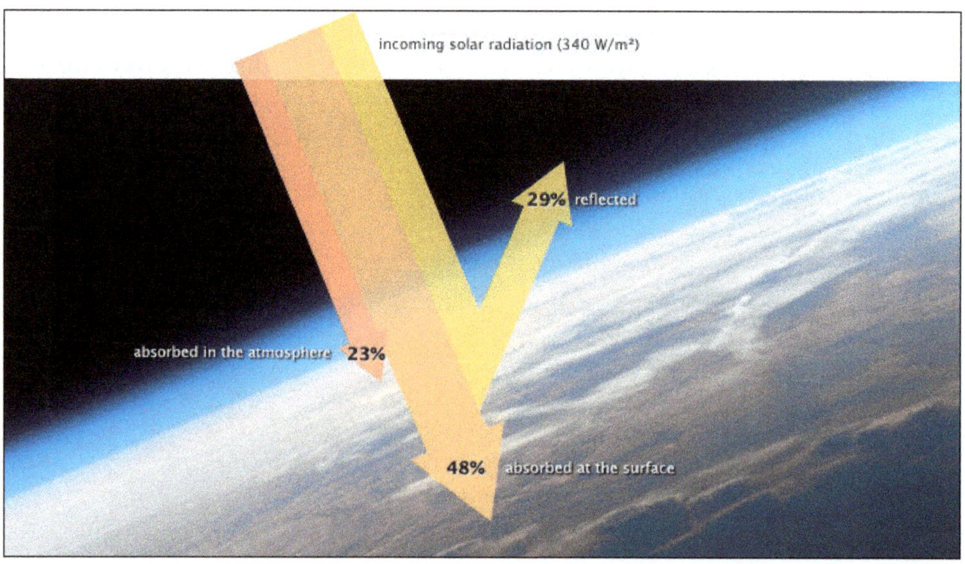

Of the 340 watts per square meter of solar energy that falls on the Earth, 29% is reflected back into space, primarily by clouds, but also by other bright surfaces and the atmosphere itself. About 23% of incoming energy is absorbed in the atmosphere by atmospheric gases, dust, and other particles. The remaining 48% is absorbed at the surface.

When matter absorbs energy, the atoms and molecules that make up the material become excited; they move around more quickly. The increased movement raises the material's temperature. If matter could only absorb energy, then the temperature of the Earth would be like the water level in a sink with no drain where the faucet runs continuously.

Temperature doesn't infinitely rise, however, because atoms and molecules on Earth are not just absorbing sunlight, they are also radiating thermal infrared energy (heat). The amount of heat a surface radiates is proportional to the fourth power of its temperature. If temperature doubles, radiated energy increases by a factor of 16. If the temperature of the Earth rises, the planet rapidly emits an increasing amount of heat to space. This large increase in heat loss in response to a relatively smaller increase in temperature—referred to as radiative cooling—is the primary mechanism that prevents runaway heating on Earth.

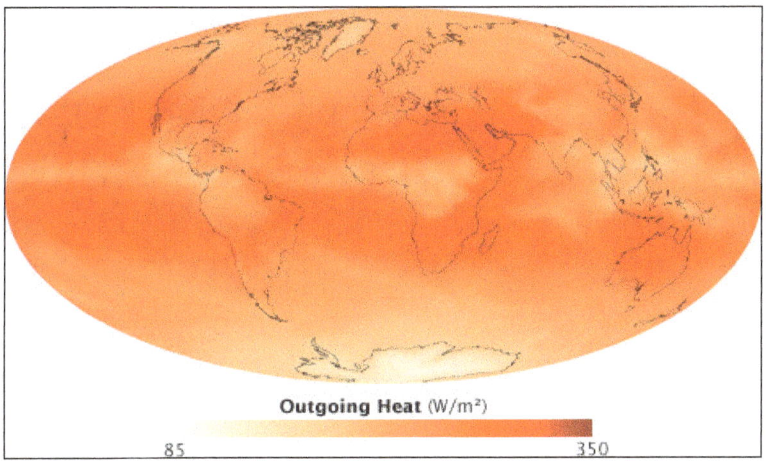

Absorbed sunlight is balanced by heat radiated from Earth's surface and atmosphere. This satellite map shows the distribution of thermal infrared radiation emitted by Earth in September 2008. Most heat escaped from areas just north and south of the equator, where the surface was warm, but there were few clouds. Along the equator, persistent clouds prevented heat from escaping. Likewise, the cold poles radiated little heat.

The atmosphere and the surface of the Earth together absorb 71 percent of incoming solar radiation, so together, they must radiate that much energy back to space for the planet's average temperature to remain stable. However, the relative contribution of the atmosphere and the surface to each process (absorbing sunlight versus radiating heat) is asymmetric. The atmosphere absorbs 23 percent of incoming sunlight while the surface absorbs 48. The atmosphere radiates heat equivalent to 59 percent of incoming sunlight; the surface radiates only 12 percent. In other words, most solar heating happens at the surface, while most radiative cooling happens in the atmosphere.

Surface Energy Budget

To understand how the Earth's climate system balances the energy budget, we have to consider processes occurring at the three levels: the surface of the Earth, where most solar heating takes place; the edge of Earth's atmosphere, where sunlight enters the system; and the atmosphere in between. At each level, the amount of incoming and outgoing energy, or net flux, must be equal.

Remember that about 29 percent of incoming sunlight is reflected back to space by bright particles in the atmosphere or bright ground surfaces, which leaves about 71 percent to be absorbed by the atmosphere (23 percent) and the land (48 percent). For the energy budget at Earth's surface to balance, processes on the ground must get rid of the 48 percent of incoming solar energy that the ocean and land surfaces absorb. Energy leaves the surface through three processes: evaporation, convection, and emission of thermal infrared energy.

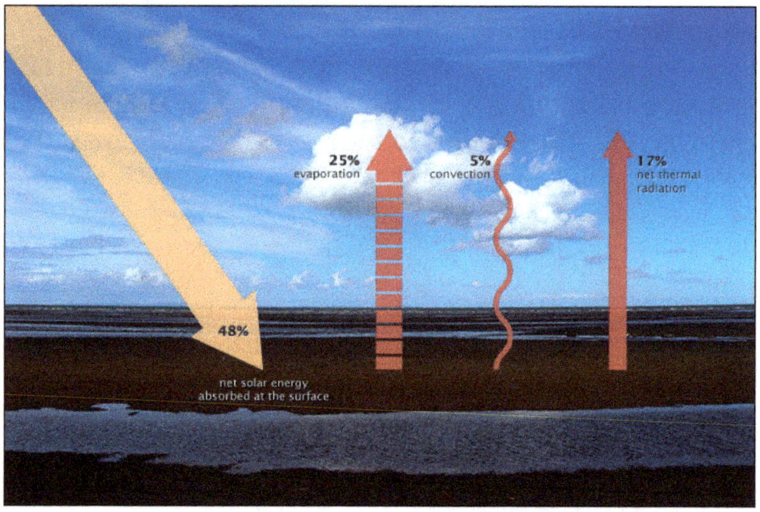

The surface absorbs about 48% of incoming sunlight. Three processes remove an equivalent amount of energy from the Earth's surface: evaporation (25%), convection (5%), and thermal infrared radiation, or heat (net 17%).

About 25 percent of incoming solar energy leaves the surface through evaporation. Liquid water molecules absorb incoming solar energy, and they change phase from liquid to gas. The heat energy that it took to evaporate the water is latent in the random motions of the water vapor molecules as they spread through the atmosphere. When the water vapor molecules condense back into rain, the latent heat is released to the surrounding atmosphere. Evaporation from tropical oceans and the subsequent release of latent heat are the primary drivers of the atmospheric heat engine.

Towers of cumulus clouds transport energy away from the surface of the Earth. Solar heating drives evaporation. Warm, moist air becomes buoyant and rises, moving energy from the surface high into the atmosphere. Energy is released back into the atmosphere when the water vapor condenses into liquid water or freezes into ice crystals.

An additional 5 percent of incoming solar energy leaves the surface through convection. Air in direct contact with the sun-warmed ground becomes warm and buoyant. In general, the atmosphere is warmer near the surface and colder at higher altitudes, and under these conditions, warm air rises, shuttling heat away from the surface.

Finally, a net of about 17 percent of incoming solar energy leaves the surface as thermal infrared energy (heat) radiated by atoms and molecules on the surface. This net upward flux results from two large but opposing fluxes: heat flowing upward from the surface to the atmosphere (117%) and heat flowing downward from the atmosphere to the ground (100%). Remember that the peak wavelength of energy a surface radiates is based on its temperature. The Sun's peak radiation is at visible and near-infrared wavelengths. The Earth's surface is much cooler, only about 15 degrees Celsius on average. The peak radiation from the surface is at thermal infrared wavelengths around 12.5 micrometers.

The Atmosphere's Energy Budget

Just as the incoming and outgoing energy at the Earth's surface must balance, the flow of energy into the atmosphere must be balanced by an equal flow of energy out of the atmosphere and back to space. Satellite measurements indicate that the atmosphere radiates thermal infrared energy equivalent to 59 percent of the incoming solar energy. If the atmosphere is radiating this much, it must be absorbing that much. Where does that energy come from?

Clouds, aerosols, water vapor, and ozone directly absorb 23 percent of incoming solar energy. Evaporation and convection transfer 25 and 5 percent of incoming solar energy from the surface to the atmosphere. These three processes transfer the equivalent of 53 percent of the incoming solar energy to the atmosphere. If total inflow of energy must match the outgoing thermal infrared observed at the top of the atmosphere, where does the remaining fraction (about 5-6 percent) come from? The remaining energy comes from the Earth's surface.

The Natural Greenhouse Effect

Just as the major atmospheric gases (oxygen and nitrogen) are transparent to incoming sunlight, they are also transparent to outgoing thermal infrared. However, water vapor, carbon dioxide, methane, and other trace gases are opaque to many wavelengths of thermal infrared energy. Remember that the surface radiates the net equivalent of 17 percent of incoming solar energy as thermal infrared. However, the amount that directly escapes to space is only about 12 percent of incoming solar energy. The remaining fraction—a net 5-6 percent of incoming solar energy—is transferred to the atmosphere when greenhouse gas molecules absorb thermal infrared energy radiated by the surface.

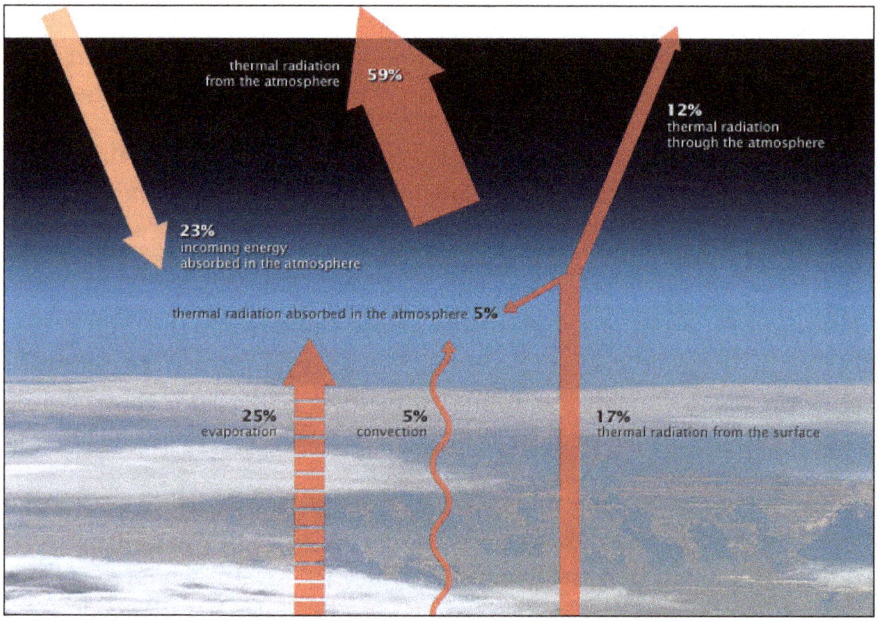

The atmosphere radiates the equivalent of 59% of incoming sunlight back to space as thermal infrared energy, or heat. Where does the atmosphere get its energy? The atmosphere directly absorbs about 23% of incoming sunlight, and the remaining energy is transferred from the Earth's surface by evaporation (25%), convection (5%), and thermal infrared radiation (a net of 5-6%). The remaining thermal infrared energy from the surface (12%) passes through the atmosphere and escapes to space.

When greenhouse gas molecules absorb thermal infrared energy, their temperature rises. Like coals from a fire that are warm but not glowing, greenhouse gases then radiate an increased amount of thermal infrared energy in all directions. Heat radiated upward continues to encounter greenhouse gas molecules; those molecules absorb the heat, their temperature rises, and the amount of heat they radiate increases. At an altitude of roughly 5-6 kilometers, the concentration of greenhouse gases in the overlying atmosphere is so small that heat can radiate freely to space.

Because greenhouse gas molecules radiate heat in all directions, some of it spreads downward and ultimately comes back into contact with the Earth's surface, where it is absorbed. The temperature of the surface becomes warmer than it would be if it were heated only by direct solar heating. This supplemental heating of the Earth's surface by the atmosphere is the natural greenhouse effect.

Effect on Surface Temperature

The natural greenhouse effect raises the Earth's surface temperature to about 15 degrees Celsius on average—more than 30 degrees warmer than it would be if it didn't have an atmosphere. The amount of heat radiated from the atmosphere to the surface (sometimes called "back radiation") is equivalent to 100 percent of the incoming solar energy. The Earth's surface responds to the "extra" (on top of direct solar heating) energy by raising its temperature.

On average, 340 watts per square meter of solar energy arrives at the top of the atmosphere. Earth returns an equal amount of energy back to space by reflecting some incoming light and by radiating heat (thermal infrared energy). Most solar energy is absorbed at the surface, while most heat is radiated back to space by the atmosphere. Earth's average surface temperature is maintained by two large, opposing energy fluxes between the atmosphere and the ground (right)—the greenhouse effect.

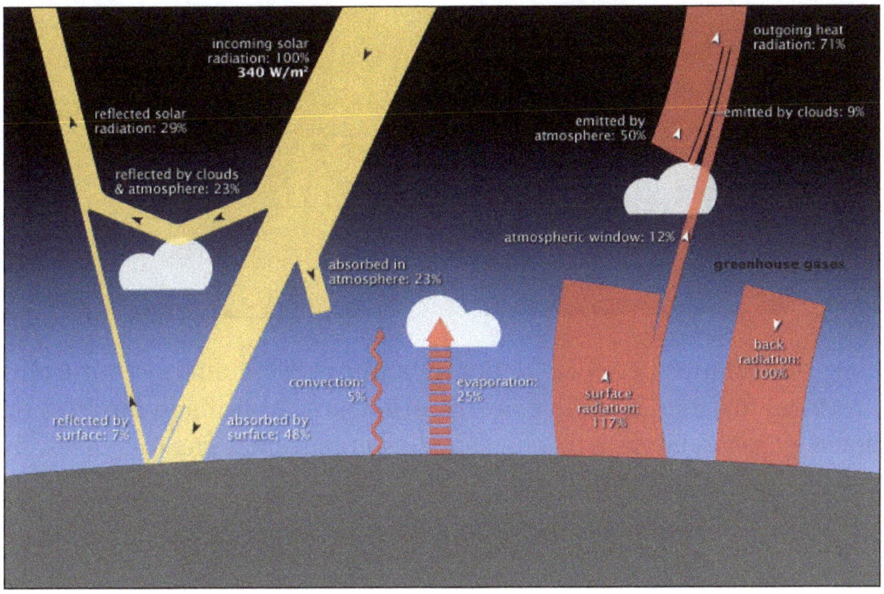

Why doesn't the natural greenhouse effect cause a runaway increase in surface temperature? Remember that the amount of energy a surface radiates always increases faster than its temperature rises—outgoing energy increases with the fourth power of temperature. As solar heating and "back radiation" from the atmosphere raise the surface temperature, the surface simultaneously releases an increasing amount of heat—equivalent to about 117 percent of incoming solar energy. The net upward heat flow, then, is equivalent to 17 percent of incoming sunlight (117 percent up minus 100 percent down).

Some of the heat escapes directly to space, and the rest is transferred to higher and higher levels of the atmosphere, until the energy leaving the top of the atmosphere matches the amount of incoming solar energy. Because the maximum possible amount of incoming sunlight is fixed by the solar constant (which depends only on Earth's distance from the Sun and very small variations during the solar cycle), the natural greenhouse effect does not cause a runaway increase in surface temperature on Earth.

Climate Forcings and Global Warming

Any changes to the Earth's climate system that affect how much energy enters or leaves the system alters Earth's radiative equilibrium and can force temperatures to rise or fall. These destabilizing influences are called climate forcings. Natural climate forcings include changes in the Sun's brightness, Milankovitch cycles (small variations in the shape of Earth's orbit and its axis of rotation that occur over thousands of years), and large volcanic eruptions that inject light-reflecting particles as high as the stratosphere. Manmade forcings include particle pollution (aerosols), which absorb and reflect incoming sunlight; deforestation, which changes how the surface reflects and absorbs sunlight; and the rising concentration of atmospheric carbon dioxide and other greenhouse gases, which decrease heat radiated to space. A forcing can trigger feedbacks that intensify or weaken the original forcing. The loss of ice at the poles, which makes them less reflective, is an example of a feedback.

Things that change the balance between incoming and outgoing energy in the climate system are called forcings. Natural forcings include volcanic eruptions. Manmade forcings include air pollution and greenhouse gases. A climate forcing, such as greenhouse gas increases, may trigger feedbacks like the loss of sunlight-reflecting ice.

Carbon dioxide forces the Earth's energy budget out of balance by absorbing thermal infrared energy (heat) radiated by the surface. It absorbs thermal infrared energy with wavelengths in a part of the energy spectrum that other gases, such as water vapor, do not. Although water vapor is a powerful absorber of many wavelengths of thermal infrared energy, it is almost transparent to others. The transparency at those wavelengths is like a window the atmosphere leaves open for radiative cooling of the Earth's surface. The most important of these "water vapor windows" is for thermal infrared with wavelengths centered around 10 micrometers. (The maximum transparency occurs at 10 micrometers, but partial transparency occurs for wavelengths between about 8 and about 14 micrometers).

Carbon dioxide is a very strong absorber of thermal infrared energy with wavelengths longer than 12-13 micrometers, which means that increasing concentrations of carbon dioxide partially "close" the atmospheric window. In other words, wavelengths of outgoing thermal infrared energy that our atmosphere's most abundant greenhouse gas—water vapor—would have let escape to space are instead absorbed by carbon dioxide.

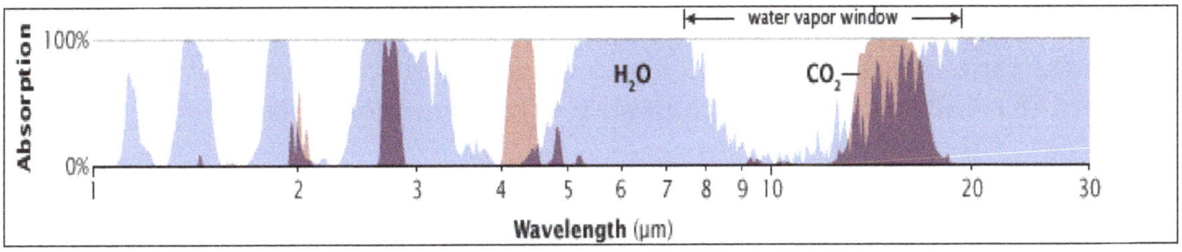

All atmospheric gases have a unique pattern of energy absorption: they absorb some wavelengths of energy but are transparent to others. The absorption patterns of water vapor (blue peaks) and carbon dioxide (pink peaks) overlap in some wavelengths. Carbon dioxide is not as strong a greenhouse gas as water vapor, but it absorbs energy in wavelengths (12-15 micrometers) that water vapor does not, partially closing the "window" through which heat radiated by the surface would normally escape to space.

The absorption of outgoing thermal infrared by carbon dioxide means that Earth still absorbs about 70 percent of the incoming solar energy, but an equivalent amount of heat is no longer leaving. The exact amount of the energy imbalance is very hard to measure, but it appears to be a little over 0.8 watts per square meter. The imbalance is inferred from a combination of measurements, including satellite and ocean-based observations of sea level rise and warming.

When a forcing like increasing greenhouse gas concentrations bumps the energy budget out of balance, it doesn't change the global average surface temperature instantaneously. It may take years or even decades for the full impact of a forcing to be felt. This lag between when an imbalance occurs and when the impact on surface temperature becomes fully apparent is mostly because of the immense heat capacity of the global ocean. The heat capacity of the oceans gives the climate a thermal inertia that can make surface warming or cooling more gradual, but it can't stop a change from occurring.

Environmental Chemistry

Earth is composed of numerous minute substances and particles which are involved in reactions leading to the formation of new ones. Hence, the earth is said to be a closed system and the energy comes and leaves the planet, of which most of the mass stays here. This means that all the elements on this planet are continuously recycling within the environment.

For example, consider free oxygen molecules that were floating in the atmosphere yesterday and it might be the part of someone's hamburger the next day. It's all up to the chemists to study these cycles and to watch their movements.

Hence, the Environmental Chemistry can be defined as The field of chemistry that deals with the study of reactions, sources, transport, effects, along with the fates of all the chemical species present in the soil, water, and the air environments, and also the effects of technology thereon.

- Environmental chemistry is the scientific study of the biochemical and chemical phenomena that occur in natural places.
- Environmental chemistry is a study that is more than air, water, soil, and chemicals. This field uses various techniques of biology, maths, genetics, engineering, hydrology, toxicology, etc. that will help to fetch an answer to all the questions related to the environment.
- Environmental chemistry also contains aspects of analytical chemistry, physical chemistry, organic chemistry & inorganic chemistry as well as more diverse areas, such as epidemiology, public health, biochemistry, biology, & toxicology.
- Environmental chemists are responsible for finding how the unpolluted environment functions and to find ways of sustainable development which do not harm the environment.

One component which a lot of environmental chemists come across, are contaminants and a good portion of our environment is a victim of contamination.

Environmental Chemistry Facts

Aluminium can be continually recycled, as in forever. Recycling one aluminium can save enough energy to run at least three hours of our TV's. Every year 80 trillion cans of aluminium are used by humans.

The chemical industry is usually associated with fluxes of polluted waste. There are, however, many other sources of water chemical pollution, including transport, agriculture, power plants and household chemicals like detergents! In fact, the chemical industry's waste streams are now strictly controlled and treated before they are released into the environment.

Chemical weathering means a chemical transformation or degradation, which usually occurs when the chemical is released into the environment. In the air, weathering is due to a combination of processes that all results in reducing the air concentration of pollutants.

Chemical water contamination can cause severe long-term problems. The consequences of acute chemical emissions on air, however, are typically short-term-but they may still be devastating.

Contamination

Contamination of the environment takes place when there are unwanted chemical substances present in the atmosphere. In fact, environmental chemistry mainly deals with the study of these unwanted substances and their effects Some contaminants of the environment are mentioned below:

- Small contaminants: Small contaminants are those metal particles that can be easily measured and controlled. These particulate contaminants can be monitored by the process of abrasion, fatigue and silting.
- Chemical Contaminants: These are the pollutants that are produced by the process of chemical reactions. They contaminate the water and soil, and those are arsenic, nitrate, fluoride, manganese atom, iron, etc.
- Gaseous contamination: This type of pollution occurs through pollution of the gaseous atmosphere by the components such as sulphur, chlorine, bromine, etc.
- Microbial contamination: This is known to be one of the dangerous types of contamination as it does not only affect the environment but also the living system within it. The cause for this is microbes such as the yeast, bacteria, mould, fungi, protozoa, virus, or the toxins and by-products.

Applications of Environmental Chemistry

Some applications of environmental chemistry are mentioned below:

- Environmental Chemistry studies the risk factors of all the chemicals in-depth to get a solution for the safety purpose of the environment.
- It is applied in the study of new products and their effects on the environment.
- Environmental Chemistry is used in the method of protection of groundwater which is polluted by soil, dust, and the waste particles.
- It is useful for the protection of surface water from the contaminants through the process of sedimentation, bacteriological, and radiation.
- The quality of the soil is protected by the methods of environmental chemicals such as by the use of indicators like ecotoxicological and chemical.
- Impervious surfaces inside cities like the parking lots, rooftops & roads are prone to build up unwanted pollutants such as motor oil, gasoline, nutrients and sediment (soil), hydrocarbon compounds, metals.
- Environmental chemistry is applied in the waste management and cleaner production.

References

- Atmospheric Structure: albany.edu, Retrieved 10, January 2020
- Ozone Formation and Destruction: ssec.wisc.edu, Retrieved 04, August 2020
- A-Case-Studies-Kinetics, Physical-and-Theoretical-Chemistry, Supplemental-Modules-(Physical-and-Theoretical-Chemistry): libretexts.org, Retrieved 19, February 2020
- environmental-chemistry: byjus.com, Retrieved 03, July 2020

Geosphere and its Chemistry

The geosphere of the Earth encompasses the rocks and minerals on Earth as well as the abiotic portions of soil and the fossilized skeletons of animals. The chemical characteristics of soil are also studied within this discipline. This chapter has been carefully written to provide an easy understanding of the varied facets of geosphere and its chemistry such as internal structure of Earth and soil formation.

Geosphere

Figure below shows a picture of the Earth taken from space. What features of the Earth can you spot in Fig? Firstly, you would note that it is not a perfect sphere. It is also blue in colour. Do you notice that it has a large water body and landmass? You also know that Earth has an atmosphere, and it supports life.

Fig: Earth as seen from the space.

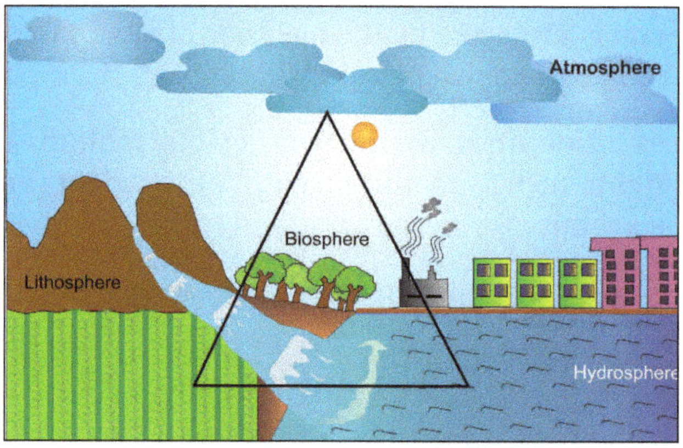

Fig: Components of the Geosphere.

Thus, we may think of the Earth as a geosphere that has four main components which are:

- Lithosphere – Solid land mass.

- Atmosphere – Envelope of air.
- Hydrosphere – Body of water.
- Biosphere – All life on Earth.

Remember, geosphere is a kind of umbrella term exhibiting all these four vital components of the planet Earth. We now briefly describe each one of these components.

The Lithosphere

Have you ever thought about how various forms of life dwell on the planet Earth? They dwell on the lithosphere which provides suitable and stable piece of landmass. This stable landmass is arranged into several topographical features such as mountains, hills, plains and plateaus etc. Which of these landforms do you dwell on? These landforms make a part of the physical environment in geosphere. The lithosphere provides suitable habitat for various forms of life. It is the source of soil, sediment, salts of the sea and nutrients. It also supports the atmosphere and stores water in its various forms. Hence, lithosphere is necessary and vital for the survival of all flora and fauna along with higher forms of life.

Lithosphere consists of the crust and upper parts of the mantle. Crust can be divided into two parts, viz., continental and oceanic crust.

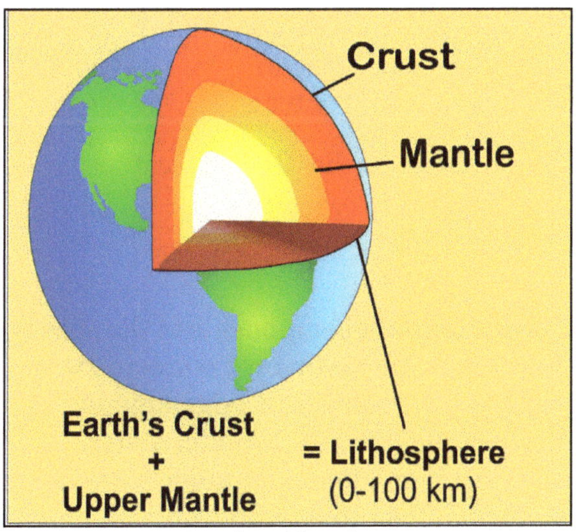

Fig: The Lithosphere.

The average thickness of the lithosphere is about 100 kilometres. However, it significantly differs between the oceans and continents. The thickness varies from only a few kilometres over the oceanic parts of the lithosphere to around 300 kilometres under the continents. Lithosphere is composed of different types of rocks such as igneous, sedimentary and metamorphic rocks.

The Atmosphere

The atmosphere is an extremely vital component of the geosphere. You may be well aware that it envelops the Earth and is made up of various gases, water and dust particles. The main gases present in the atmosphere are nitrogen and oxygen having a share of nearly 75 and 24 percent. Many other gases such as argon, CO_2, helium and neon are present in minute quantities. The amount of these gases in the atmosphere decreases as we move away from the Earth.

Almost the entire atmosphere constituting about 97 percent extends up to 30 kilometres from the Earth's surface. Its upper limit is at approximately 1000 kilometres. Try to compare these distances with the diameter of the Earth (12,800 kilometres). The space beyond this distance (about 100 kilometres) is generally referred to as extra-terrestrial space.

To understand the importance of atmosphere, think about this question. If there would be no atmosphere on the Earth, would life have been possible? The atmosphere plays an important role for the existence of various forms of life. It performs following functions:

- Provides air to breathe and water to drink.
- Protects all life forms, flora and fauna from the Sun's harmful rays.
- Protects us from the meteorites falling from space.

It is a lively sphere being always in continuous action driven by the solar energy and Earth's rotation. Thus, we cannot imagine any form of life on the Earth in the absence of this protective layer. We now understand that atmosphere is a dynamic component of the geosphere. It has two types of interactions:

- Interaction between atmosphere and the other three components of the geosphere.
- Interaction between atmosphere and outer space.

Both these types of interactions give rise to climate and weather processes.

The Hydrosphere

So far, you have learnt about two important components of the geosphere. We now discuss the hydrosphere, which is equally important. Have you ever pondered why Earth is called a blue planet? We call it a blue planet because of the presence of large amounts of water on it. Approximately, three-fourths of the surface area of the Earth lies under water. Water is present in three states gaseous (water vapour in the atmosphere), liquid (running water on the surface and underground water beneath the surface) and solid (ice caps). Water present in the Earth's surface is mainly of two kinds:

- Fresh water (about 3%) is stored in the glaciers, lakes, streams and underground.
- Salt water (about 97%), stored in seas and oceans.

Refer to below figure, which shows the distribution of water on the Earth.

All these forms of water are dynamic in nature. Water is stored in glaciers, lakes, rivers, seas and oceans etc. The sphere which stores enormous amounts of water as ice in frozen state is known as cryosphere (subcomponent of hydrosphere). The stored water in its liquid state evaporates into the atmosphere and, falls back in the form of precipitation (rain, snow and hailstorm etc.) on the

Earth's surface. Then, it flows back to the various water bodies. This process of evaporation and precipitation is continuous. It is called the 'hydrological cycle'.

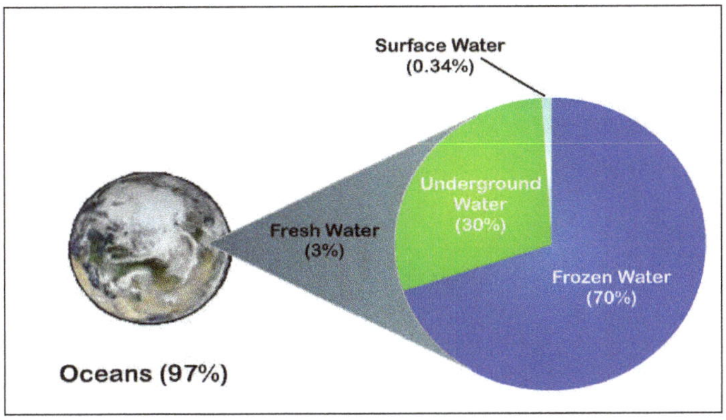

Water in oceans, seas, rivers and streams is also responsible for changing and creating various landforms. For example, the rocks on sea beaches are constantly sculptured by water waves resulting into formation of various landforms like cliffs and sea caves etc.

As you know, water is essential for various forms of life. Even our body is composed of nearly 75% water. The hydrosphere is a source of the huge amounts of water needed for agriculture, industry and other economic activities. You will agree that modern society would not exist in its absence.

The Biosphere

So far, you have learnt that lithosphere, atmosphere and hydrosphere constitutes an integral part of the geosphere. These three are inorganic or non-living components. Now, you will learn about the organic or living component called the biosphere. This is the fourth most important and lively component. Therefore, it also includes those parts of the lithosphere, atmosphere and hydrosphere that provide the physical environment.

Can you imagine how the life forms evolved over the planet Earth? All life forms have evolved in a narrow zone in the biosphere. It includes plants, animals and all other living things. You may think that these life forms are self-dependent for their needs of food and water. It is not so. They are not independent as they share complex relationship with the other three components of the geosphere.

For example, the oxygen we breathe is provided by the atmosphere; the water we drink comes from the hydrosphere and the geographic area where we live is on the lithosphere. Thus, together the other three components provide the physical environment that supports all life forms. You can state that biosphere includes:

- All life forms on the Earth.
- Those parts of the lithosphere, atmosphere and hydrosphere that provide the physical environment to support life.

You may have noticed that most of life forms are confined to the Earth's surface. It includes all flora and fauna along with human beings. Life is mostly confined in a narrow zone of the geosphere. It is concentrated between the lowest layers of the atmosphere and the ocean bottoms in the biosphere. More than three-fourths of the all life forms are concentrated here. However, there is an exception to this.

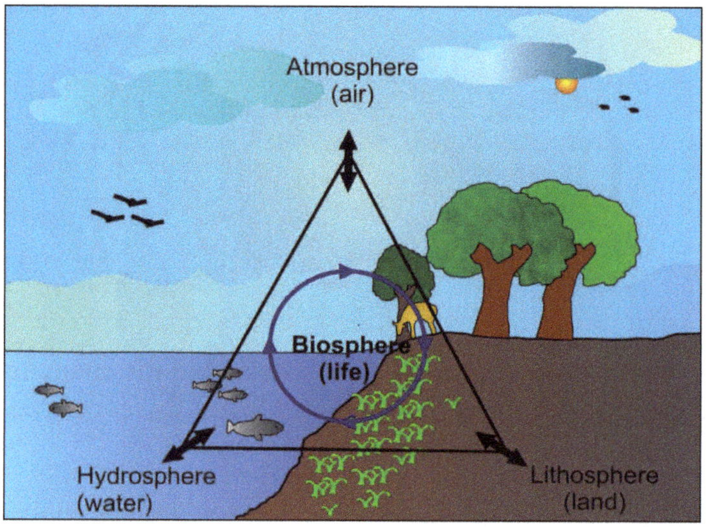

Can you imagine that life forms also develop beyond this narrow zone? Well, it is true. Do you know examples of such life forms? The coral reef that develops in the deep ocean floor is one such example. Another example is bacteria that thrive and live on deep parts of the rock layers beneath the Earth's surface up to 4 kilometres.

We burn fossil fuels (coal, oil and gas) for different purposes such as cooking, transportation and running of factories and industrial units. Burning of fossil fuels adds up carbon dioxide and other gases such like sulphur dioxide and carbon monoxide into the narrow zone of biosphere.

This additional amount of carbon dioxide may disturb the natural balance in rest of the three components of the geosphere. Such kinds of disturbance may generate harmful effects over the only liveable planet Earth. It may lead to disasters such as climate change, air and water pollution etc. to name a few with which you are familiar.

One of the best examples of man-made 'Biosphere' in the world is that of 'Biosphere 2' located in the University of Arizona, USA. It provides one of the world's most unique facilities dedicated to the research and understanding of global scientific issues.

You can say that all the four components of geosphere namely lithosphere, atmosphere, hydrosphere and biosphere are linked to each other. In other words, all are dependent on each other for the exchange of energy and material. They co-exist and one cannot function without the help of other. You will learn about this dependency and exchange of energy and matter in the next section on ecosystems.

Thus, the biosphere quintessentially regularly alters the character and form of lithosphere, atmosphere and hydrosphere. It is the only lively component that renders a definite purpose to the other three non-lively components.

Internal Structure of Earth

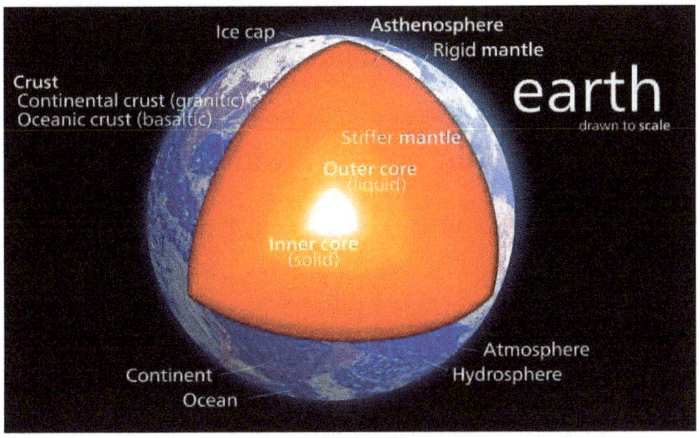

Fig: Earth's Layers.

- The interior of the earth is made up of several concentric layers of which the crust, the mantle, the outer core and the inner core are significant because of their unique physical and chemical properties.
- The crust is a silicate solid, the mantle is a viscous molten rock, the outer core is a viscous liquid, and the inner core is a dense solid.

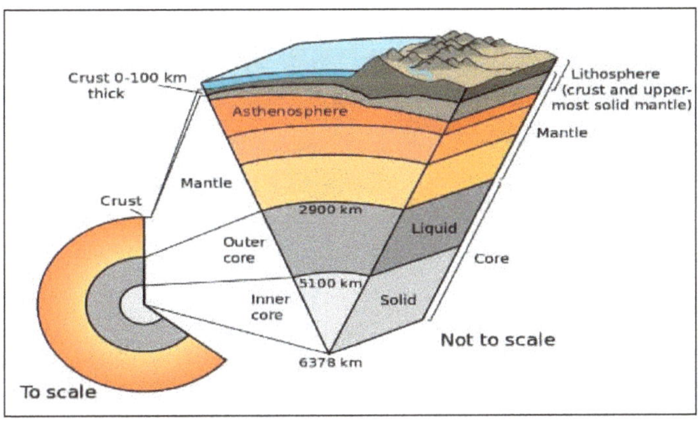

- Mechanically, the earth's layers can be divided into lithosphere, asthenosphere, mesospheric mantle (part of the Earth's mantle below the lithosphere and the asthenosphere), outer core, and inner core.
- Chemically, Earth can be divided into the crust, upper mantle, lower mantle, outer core, and inner core.

The Crust

- The crust is the outermost layer of the earth making up 0.5-1.0 per cent of the earth's volume and less than 1 per cent of Earth's mass.
- Density increases with depth, and the average density is about 2.7 g/cm^3 (average density of the earth is 5.51 g/cm^3).

- The thickness of the crust varies in the range of range of 5-30 km in case of the oceanic crust and as 50-70 km in case of the continental crust.
- The continental crust can be thicker than 70 km in the areas of major mountain systems. It is as much as 70-100 km thick in the Himalayan region.
- The temperature of the crust increases with depth, reaching values typically in the range from about 200 °C to 400 °C at the boundary with the underlying mantle.
- The temperature increases by as much as 30 °C for every kilometer in the upper part of the crust.
- The outer covering of the crust is of sedimentary material and below that lie crystalline, igneous and metamorphic rocks which are acidic in nature.
- The lower layer of the crust consists of basaltic and ultra-basic rocks.
- The continents are composed of lighter silicates — silica + aluminium (also called sial) while the oceans have the heavier silicates — silica + magnesium (also called sima).
- The continental crust is composed of lighter (felsic) sodium potassium aluminium silicate rocks, like granite.
- The oceanic crust, on the other hand, is composed of dense (mafic) iron magnesium silicate igneous rocks, like basalt.

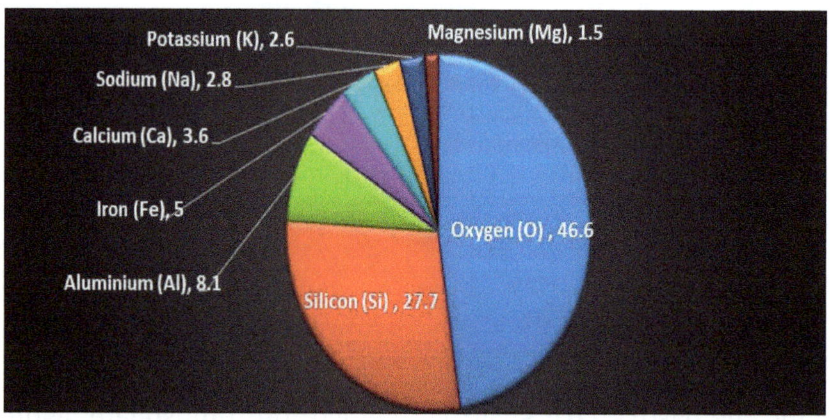

In geology, felsic refers to igneous rocks that are relatively rich in elements that form feldspar and quartz. It is contrasted with mafic rocks, which are relatively richer in magnesium and iron.

Felsic refers to rocks which are enriched in the lighter elements such as silicon, oxygen, aluminium, sodium, and potassium.

Table: Most Abundant Elements of the Earth's Crust.

	Element	Approximate % by weight
1	Oxygen (O)	46.6
2	Silicon (Si)	27.7

3	Aluminium (Al)	8.1
4	Iron (Fe)	5.0
5	Calcium (Ca)	3.6
6	Sodium (Na)	2.8
7	Potassium (K)	2.6
8	Magnesium (Mg)	1.5

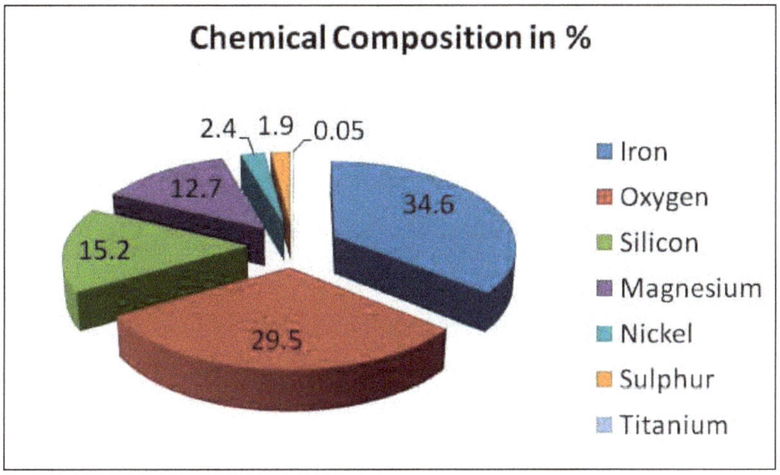

Most Abundant Elements of the Earth.

The Mohorovicic (Moho) Discontinuity

- Mohorovicic (Moho) discontinuity forms the boundary between the crust and the asthenosphere (upper reaches of the mantle) where there is a discontinuity in the seismic velocity.
- It occurs at an average depth of about 8 kilometres beneath the ocean basins and 30 kilometres beneath continental surfaces.
- The cause of the Moho is thought to be a change in rock composition from rocks containing feldspar (above) to rocks that contain no feldspars (below).

Lithosphere

- The lithosphere is the rigid outer part of the earth with thickness varying between 10-200 km.
- It is includes the crust and the upper part of the mantle.
- The lithosphere is broken into tectonic plates (lithospheric plates), and the movement of these tectonic plates cause large-scale changes in the earth's geological structure (folding, faulting).
- The source of heat that drives plate tectonics is the primordial heat left over from the planet's formation as well as the radioactive decay of uranium, thorium, and potassium in Earth's crust and mantle.

The Mantle

- It forms about 83 per cent of the earth's volume and holds 67% of the earth's mass.
- It extends from Moho's discontinuity to a depth of 2,900 km.
- The density of the upper mantle varies between 2.9 g/cm^3 and 3.3 g/cm^3.
- The lower mantle extends beyond the asthenosphere. It is in a solid state.
- The density ranges from 3.3 g/cm^3 to 5.7 g/cm^3 in the lower mantle.
- The mantle is composed of silicate rocks that are rich in iron and magnesium relative to the overlying crust.
- Regarding its constituent elements, the mantle is made up of 45% oxygen, 21% silicon, and 23% magnesium (OSM).
- In the mantle, temperatures range from approximately 200 °C at the upper boundary with the crust to approximately 4,000 °C at the core-mantle boundary.
- Because of the temperature difference, there is a convective material circulation in the mantle (although solid, the high temperatures within the mantle cause the silicate material to be sufficiently ductile).
- Convection of the mantle is expressed at the surface through the motions of tectonic plates.
- High-pressure conditions ought to inhibit seismicity in the mantle. However, in subduction zones, earthquakes are observed down to 670 km (420 mi).

Asthenosphere

- The upper portion of the mantle is called as asthenosphere (astheno means weak).
- It lies just below the lithosphere extending up to 80-200 km.
- It is highly viscous, mechanically weak and ductile and its density is higher than that of the crust.
- These properties of the asthenosphere aid in plate tectonic movement and isostatic adjustments (the elevated part at one part of the crust area is counterbalanced by a depressed part at another).
- It is the main source of magma that finds its way to the surface during volcanic eruptions.

The Outer Core

- The outer core, surrounding the inner core, lies between 2900 km and 5100 km below the earth's surface.
- The outer core is composed of iron mixed with nickel (nife) and trace amounts of lighter elements.

- The outer core is not under enough pressure to be solid, so it is liquid even though it has a composition similar to the inner core.
- The density of the outer core ranges from 9.9 g/cm³ to 12.2 g/cm³.
- The temperature of the outer core ranges from 4400 °C in the outer regions to 6000 °C near the inner core.
- Dynamo theory suggests that convection in the outer core, combined with the Coriolis effect, gives rise to Earth's magnetic field.

The Inner Core

- The inner core extends from the centre of the earth to 5100 km below the earth's surface.
- The inner core is generally believed to be composed primarily of iron (80%) and some nickel (nife).
- Since this layer can transmit shear waves (transverse seismic waves), it is solid. (When P-waves strike the outer core – inner core boundary, they give rise to S-waves).
- Earth's inner core rotates slightly faster relative to the rotation of the surface.
- The solid inner core is too hot to hold a permanent magnetic field.
- The density of the inner core ranges from 12.6 g/cm³ to 13 g/cm³.
- The core (inner core and the outer core) accounts for just about 16 per cent of the earth's volume but 33% of earth's mass.
- Scientists have determined the temperature near the Earth's centre to be 6000° C, 1000° C hotter than previously thought.
- At 6000°C, this iron core is as hot as the Sun's surface, but the crushing pressure caused by gravity prevents it from becoming liquid.

Remember when ambient pressure increases the melting point of solid increases, and vice versa. One exception is Ice. In the case of ice increase in ambient pressure will lower its melting point.

Soil Formation

Pedogenesis or soil evolution (formation) is the process by which soil is formed.

Origin of Soils

The force of wind, water or glaciers might have transported the soil to some other place. In addition to the soil parent material, origin of soil is also dependent upon other prevailing processes affecting soil formation. Climatic conditions are important factors affecting both the form and rate of physical and chemical weathering of the parent material. The formation of soils can be seen as a combination of the products of weathering, of structural development of the soil, of differentiation of that structure into horizons or layers, and lastly of its movement or translocation.

Factors of Soil Formation

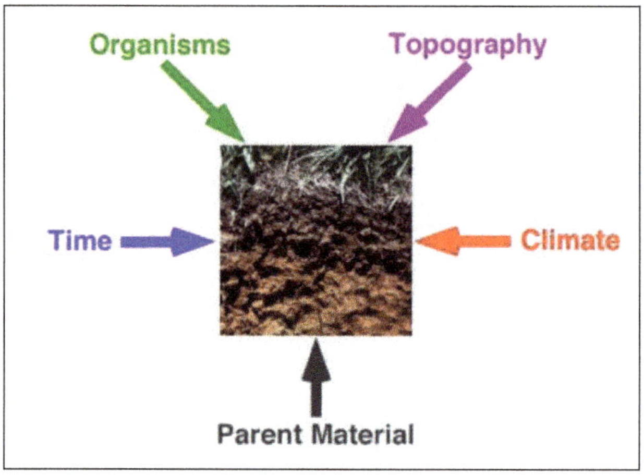

Factors Affecting the Formation of Soil

The Jenny Equation

Soil scientist, Hans Jenny has suggested that type of soil found on any site is dependent upon the interaction of five factors.

Jenny's state factor equation for soil genesis:

- S = f (C, O, R, P, T,...)
- C = Climate
- O = Organisms
- R = Relief (topography)
- P = Parent Material
- T = Time
- ...= other unspecified factors

Any of the physical, chemical or biological processes taking place in the soil as a result of these factors are called pedogenic processes.

State Factor Analysis

Folger's concept of sediment genesis:

- Se = f (G, H, B) (G, H, B)

- Se = sediment characteristics
- G = source geology
- H = hydrology (flow regime)
- B = bathymetry

Time

Soil formation takes several hundreds to thousands of years to undergo significant changes. Soils can take many years to form. Younger soils have some characteristics from their parent material, but as they age, the addition of organic matter, exposure to moisture and other environmental factors may change its features. With time, they settle and are buried deeper below the surface, taking time to transform. Eventually they may change from one soil type to another. Most of the soils of the world have taken more than 10,000 years to form the current state of soils.

Parental Material

The parental material determines the mineralogical composition and widely contributes to the chemical and physical characteristics of the soil. The type of parental material also determines the rate at which soil forming processes occurs. Different places have different soils based on the parent materials such as marine area peaty soils due to the marine organic deposits and flood plain clay soils due to alluvial deposits.

Accordingly, the complexity of soil patterns, texture, composition, and color in different areas highly depends on the physical and chemical compositions of the parent materials. Majority of loamy soil, for instance, forms as a result of thin deposits of fine grained materials that have been mixed with organic materials and other underlying materials through natural process.

Most soil parent materials were rocks at some time in their history. The minerals in rocks contribute to soil fertility and other soil properties long after the original rock is gone. Consequently, it is a valuable skill to be able to identify a few broad categories of rock. Geologists classify rocks into igneous, sedimentary and metamorphic rocks, according to their origins.

Weathering breaks up rock and minerals, changes their chemical composition and characteristics, and also synthesizes a soil from the upper part of the regolith. This unconsolidated material is act-ed on by the five soil forming factors and soil is formed. For this reason the upper part of the regolith is called parent material.

There are 5 general categories of parent material-Residual Rocks and Minerals:

- Glacial Deposits.
- Loess Deposits.

- Alluvial and Marine Deposits.
- Organic Deposits.

Two types of minerals are found in natural systems: primary and secondary.

- Whether the mineral is primary or secondary depends on the mode of formation and not on the mineral composition.
- Minerals that crystallize from cooling magma are called primary.
- Minerals that crystallize during the weathering of primary minerals are called secondary minerals.

Climate

Climate especially precipitation, temperature and frost action have a fundamental influence on the soil formation process that takes place within any given location. The prevailing climatic conditions highly determine the nature of weathering process that will take place and the rates of physical and chemical processes.

Climate directly has an effect on the kind of vegetation in an area which in turn will affect the soil formation processes related to root penetration and vegetation cover. The availability of moisture also has an influence on soil pH and the decomposition of organic matter. The accumulation and decay of organic matter also depends on humidity and temperature.

Rainfall leaches away soluble materials and iron-rich minerals from the upper soil horizons into the lower ones and evaporation brings about the accumulation of salt compounds in the surface horizons.

Cold winter temperatures give room for frost action which physically disintegrates the rocks into fragments.

The dominant climates recognized are:

- Arid climate: The precipitation here is far less than the water-need. Hence the soils remain dry for most of the time in a year.
- Humid climate: The precipitation here is much more than the water need. The excess water results in leaching of salt and bases followed by translocation of clay colloids.
- Oceanic climate: Moderate seasonal variation of rainfall and temperature.
- Mediterranean climate: The moderate precipitation. Winters and summers are dry and hot.
- Continental climate: Warm summers and extremely cool or cold winters.
- Temperate climate: Cold humid conditions with warm summers.
- Tropical and subtropical climate: Warm to hot humid with isothermal conditions in the tropical zone.

Climate affects the soil formation directly and indirectly. Directly, climate affects the soil formation by supplying water and heat to react with parent material. Indirectly, it determines the fauna and flora activities which furnish a source of energy in the form of organic matter.

This energy acts on the rocks and minerals in the form of acids, and salts are released. The indirect effects of climate on soil formation are most clearly seen in the relationship of soils to vegetation.

Precipitation

Precipitation is the most important among the climatic factors. As it percolates and moves from one part of the parent material to another. It carries with it substances in solution as well as in suspension. The substances so carried are re deposited in another part or completely removed from the material through percolation when the soil moisture at the surface evaporates causing an upward movement of water. The soluble substances move with it and are translocated to the upper layer. Thus rainfall brings about a redistribution of substances both soluble as well as in suspension in soil body.

Temperature

- Temperature is another climatic agent influencing the process of soil formation.
- High temperature hinders the process of leaching and causes an upward movement of soluble salts.
- High temperature favors rapid decomposition of organic matter and increase microbial activities in soil while low temperatures induce leaching by reducing evaporation and there by favour the accumulation of organic matter by slowing down the process of decomposition. Temperature thus controls the rate of chemical and biological reactions taking place in the parent material.

Topography and Relief

Topography is the configuration of a land surface and the relations among its man-made and natural features. Typically, it is the shape of the land surface and its position as well as slope on the landscape. So, these characteristics all together highly determine the types of soils formed within a region. In most regions, soils formed from similar parent materials under the same climatic conditions present differences due to their position on the landscape. The differences are primarily because of varying drainage conditions owed to the depth of the water table or surface runoff. Soils formed on sloping areas and higher elevations are by and large excessively drained.

Soil Formation on Flat to Almost Flat Position

On level topographic conditions, almost the entire water received through rainfall percolates through the soil. Under such conditions, the soils formed may be considered as representative of the regional climate. They have normal solum with distinct horizons. But vast and monotonous level land with little gradient often has impaired drainage conditions.

Soil Formation on Undulating Topography

The soils on steep slopes are generally shallow, stony and have weakly- developed profiles with less distinct horizonation. It is due to accelerated erosion, which removes surface material before it has the time to develop.

Reduced percolation of water through soil is because of surface runoff, and lack of water for the growth of plants, which are responsible for checking of erosion and promote soil formation.

Soil Formation in Depression

The depression areas in semi-arid and sub humid regions reflect more moist conditions than actually observed on level topographic positions due to the additional water received as runoff. Such conditions (as in the Tarai region of the Uttar Pradesh) favour more vegetative growth and slower rate of decay of organic remains. This results in the formation of comparatively dark- coloured soils rich in organic matter.

Organisms

All living organisms play an active role in the soil formation processes. Organisms including fungi, bacteria, animals, humans, and vegetations are the major determinants and they impact on the physical and chemical environments of the soils.

Micro-organisms encourage acidic conditions which change the soils chemistry and eventually determine the kind of soil formation process that occur.

Microbial activities also decompose organic matter and recycle them in the soil. Larger animals including burrowing animals and earthworms mix the soil and alter its physical characteristics. Man's activates have as well made tremendous changes to the natural soils. Through cultivation, construction, and addition of fertilizer and lime has altered the physical and chemical properties of the soil. Filling, mining, and artificial drainage have altered the natural soil environments thereby negatively and positively affecting the process of soil formation.

Four Soil Forming Processes

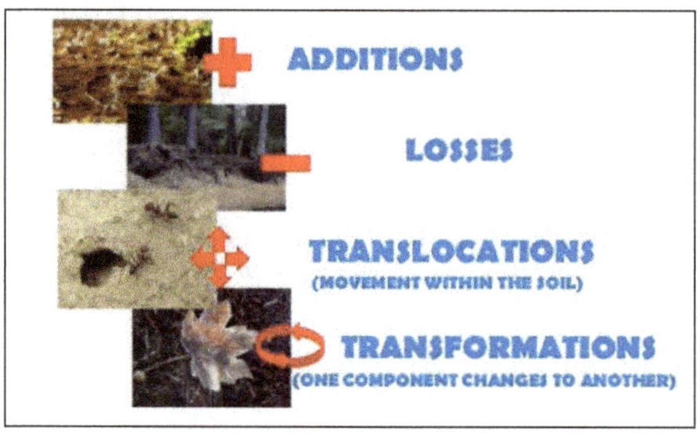

4 Basic Processes in the soil.

- Additions: Materials added to the soil, such as decomposing vegetation and organisms (organic matter--OM), or new mineral materials deposited by wind or water.

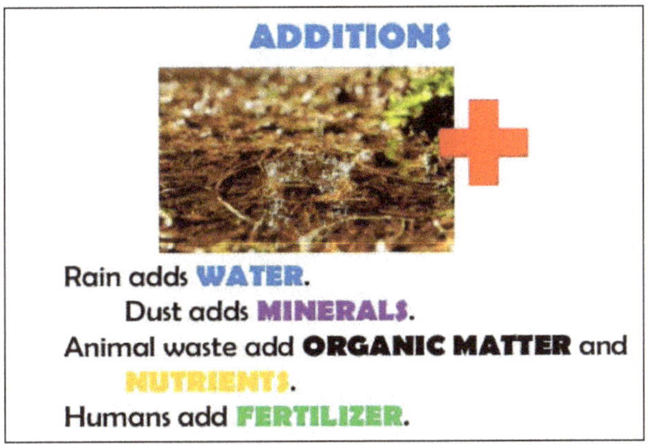

- Losses: Through the movement of wind or water, or uptake by plants, soil particles (sand, silt, clay, and OM) or chemical compounds can be eroded, leached, or harvested from the soil, altering the chemical and physical makeup of the soil.

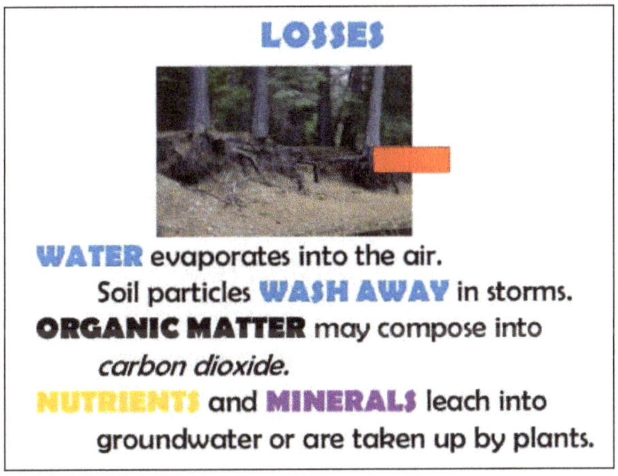

- Transformations: The chemical weathering of sand and formation of clay minerals, transformation of coarse OM into decay resistant organic compounds (humus).

- Translocations: Movement of soil constituents (organic or mineral) within the profile and/or between horizons. Over time, this process is one of the more visibly noticeable as alterations in color, texture, and structure become apparent.

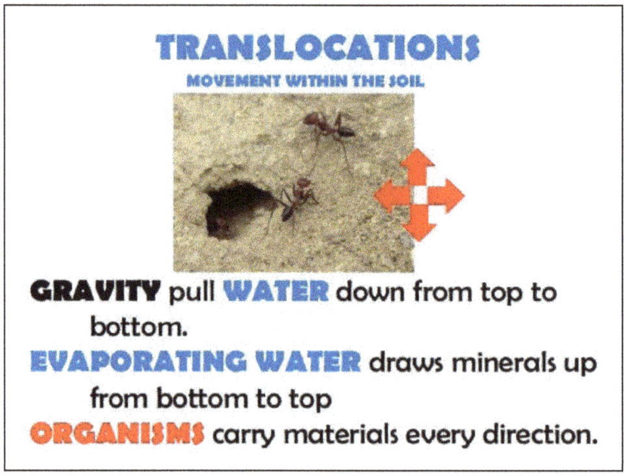

Seven Mechanisms of Soil Formation

Accumulation of Materials

Materials are added to the soil such as organic matter and decomposing materials or new mineral materials deposited by the forces of ice, water or wind and they accumulate over time. This happens in the top layer of the soil. In poorly drained soils, the organic materials accumulate since water-logging prevents it from being oxidized or broken down by soil organisms.

In well drained soils, the materials accumulate when they are held up by the root systems. Depositions by the forces of wind, water or ice equally contribute to the accumulation of new materials. Some plants with the help of bacteria fix atmospheric nitrogen and ammonia compounds into the soil as nitrates.

Leaching and Losses

Leaching is the removal of soluble components of the soil column. As water washes down through the soil it can carry away bases such as calcium, held as exchangeable ions in clay-humus complexes, as well as acidification through the substitution of hydrogen ions. Through the movement of water, wind, ice or the uptake of the accumulated materials by plants, the new particles including clay, organic matter, clay, silt or other chemical compounds are leached and eroded away or taken up from the soil by plants. As a result, the physical and chemical compositions of the new accumulated materials together with the soil parent material are altered. A prime example is the leaching out of some carbonates, magnesium and other minerals.

Transformation and Illuviation

Here the soil particles held in the suspension after the leaching such as clay are transformed after which they accumulate. Transformation is the chemical weathering of silt, sand, and the formation of clay minerals as well as the change of organic materials into decay resistant organic matter. After, the clay and other accumulated materials are washed from the upper horizons and deposited in the lower horizons. The plants and animals are also responsible for transformation of the soil by physically and chemically breaking down the materials. The soil begins to take shape on its own through transformation, which improves natural drainage and nutrient composition.

- Illuviation: Here soil particles held in suspension, such as clay, are accumulated (e.g. deposited).
- Eluviation: Here soil particles held in suspension, such as clay, are removed (e.g. washed away).

Podsolisation and Translocations

Podsolisation occurs when strongly acid soil solutions cause the breakdown of clay minerals. As a result silica, aluminium and iron form complexes with organic substances in the soil. These minerals are removed from the surface zone of the soil and can accumulate in distinct dark sub-surface layers - very evident on inspection. Upland heaths and moors often contain podsols. Podsolisation takes place when strong acidic solutions breakdown the clay minerals.

Accordingly, aluminum, silica and iron form complex materials together with organic compounds in the soil. These materials and the other accumulations are translocated within the profile and/or between the horizons. After a relatively extended time, the movements of the accumulated mineral precipitate such as iron oxides minerals are responsible for the reddish and brownish color of the different soil horizons.

Laterization

It is a pedogenic process common to soils found in tropical and subtropical environments. High temperatures and heavy precipitation result in the rapid weathering of rocks and minerals. Movements of large amounts of water through the soil cause eluviation and leaching to occur.

Calcification

It occurs when evapotranspiration exceeds precipitation causing the upward movement of dissolved alkaline salts from the groundwater. At the same time, the movement of rain water causes a downward movement of the salts. The net result is the deposition of the translocated cations in the B horizon. In some cases, these deposits can form a hard layer called caliche.

Gleying

Gleying occurs in waterlogged, anaerobic conditions when iron compounds are reduced and either removed from the soil, or segregated out as mottles or concretions in the soil. Marshy wetlands often contain gleyed soils.

Weathering Processes

Weathering is the breakdown and alteration of rocks and minerals at or near the Earth's surface into products that are more in equilibrium with the conditions found in this environment.

The products of weathering are a major source of sediments for erosion and deposition. The residue of weathering consists of chemically altered and unaltered materials. The most common unaltered residue is quartz. Many of the chemically altered products of weathering become very simple small compounds or nutrient ions. Some of the products of weathering, less resistant alumino-silicate minerals, become clay particles.

There are three broad categories of mechanisms for weathering: chemical, physical and biological.

Chemical Weathering

Chemical weathering involves the alteration of the chemical and mineralogical composition of the weathered material. A number of different processes can result in chemical weathering. The most common chemical weathering processes are hydrolysis, oxidation, reduction, hydration, carbonation, and solution.

- Hydrolysis is the weathering reaction that occurs when the two surfaces of water and compound meet.
- Oxidation is the reaction that occurs between compounds and oxygen. The net result of this reaction is the removal of one or more electrons from a compound, which causes the structure to be less rigid and increasingly unstable.
- Hydration is simply the reverse of oxidation, and is thus caused by the addition of one or more electrons producing a more stable compound.
- Carbonation is the reaction of carbonate and bicarbonate ions with minerals. The formation of carbonates usually takes place as a result of other chemical processes.

The most important factor affecting the entire chemical weathering processes is climate. Climatic conditions control the rate of weathering that takes place by regulating the catalysts of moisture and temperature.

Physical Weathering

Physical weathering is the breakdown of mineral or rock material by entirely mechanical methods brought about by a variety of causes. Upon freezing the volumetric change of water from liquid to solid is 9 %. This relatively large volumetric change upon freezing has potentially a great rupturing effect. Abrasion occurs when some force causes two rock surfaces to come together causing mechanical wearing or grinding of their surfaces. Collision between rock surfaces normally occurs through the erosional transport of material by wind, water, or ice. Alternate wetting and drying of rocks, sometimes known as slaking, can be a very important factor in weathering.

Biological Weathering

Biological weathering involves the disintegration of rock and mineral due to the chemical and/ or physical agents of an organism. The types of organisms that can cause weathering range from bacteria to plants to animals.

Biological weathering involves processes that can be either chemical or physical in character.

Soil Pollution

Soil pollution refers to the contamination of soil with anomalous concentrations of toxic substances. It is a serious environmental concern since it harbours many health hazards. For example, exposure to soil containing high concentrations of benzene increases the risk of contracting leukaemia. An image detailing the discolouration of soil due to soil pollution is provided below.

It is important to understand that all soils contain compounds that are harmful/ toxic to human beings and other living organisms. However, the concentration of such substances in unpolluted soil is low enough that they do not pose any threat to the surrounding ecosystem. When the concentration of one or more such toxic substances is high enough to cause damage to living organisms, the soil is said to be contaminated.

The root cause of soil pollution is often one of the following:

- Agriculture (excessive/improper use of pesticides).
- Excessive industrial activity.
- Poor management or inefficient disposal of waste.

The challenges faced in soil remediation (decontamination of soil) are closely related to the extent of soil pollution. The greater the contamination, the greater the requirement of resources for remediation.

What are the Pollutants that Contaminate Soil?

Some of the most hazardous soil pollutants are xenobiotics – substances that are not naturally found in nature and are synthesized by human beings. The term 'xenobiotic' has Greek roots – 'Xenos' (foreigner), and 'Bios' (life). Several xenobiotics are known to be carcinogens. An illustration detailing major soil pollutants is provided below.

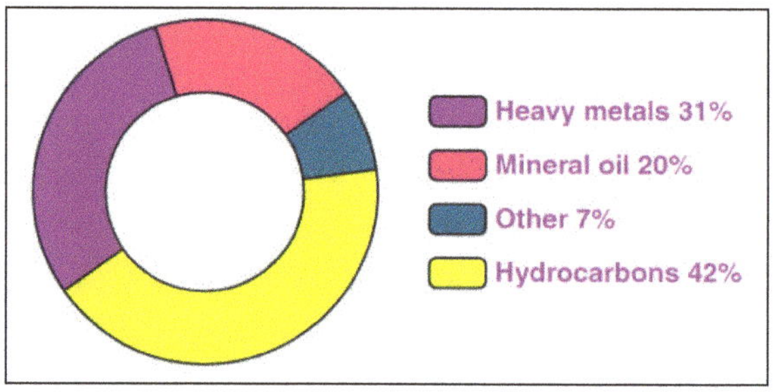

The different types of pollutants that are found in contaminated soil are listed below.

Heavy Metals

The presence of heavy metals (such as lead and mercury, in abnormally high concentrations) in soils can cause it to become highly toxic to human beings. Some metals that can be classified as soil pollutants are given in the table below:

Toxic Metals that cause soil pollution		
Arsenic	Mercury	Lead
Antimony	Zinc	Nickel
Cadmium	Selenium	Beryllium
Thallium	Chromium	Copper

These metals can originate from several sources such as mining activities, agricultural activities, electronic waste (e-waste), and medical waste.

Polycyclic Aromatic Hydrocarbons

Polycyclic aromatic hydrocarbons (often abbreviated to PAHs) are organic compounds that:

- Contain only carbon and hydrogen atoms.
- Contain more than one aromatic ring in their chemical structures.

Common examples of PAHs include naphthalene, anthracene, and phenalene. Exposure to polycyclic aromatic hydrocarbons has been linked to several forms of cancer. These organic compounds can also cause cardiovascular diseases in humans.

Soil pollution due to PAHs can be sourced to coke (coal) processing, vehicle emissions, cigarette smoke, and the extraction of shale oil.

Industrial Waste

The discharge of industrial waste into soils can result in soil pollution. Some common soil pollutants that can be sourced to industrial waste are:

- Chlorinated industrial solvents.
- Dioxins produced from the manufacture of pesticides and the incineration of waste.

- Plasticizers/dispersants.
- Polychlorinated biphenyls (PCBs).

The petroleum industry creates many petroleum hydrocarbon waste products. Some of these wastes, such as benzene and methylbenzene, are known to be carcinogenic in nature.

Pesticides

Pesticides are substances (or mixtures of substances) that are used to kill or inhibit the growth of pests. Common types of pesticides used in agriculture include:

- Herbicides: Used to kill/control weeds and other unwanted plants.
- Insecticides: Used to kill insects.
- Fungicides: Used to kill parasitic fungi or inhibit their growth.

However, the unintentional diffusion of pesticides into the environment (commonly known as 'pesticide drift') poses a variety of environmental concerns such as water pollution and soil pollution. Some important soil contaminants found in pesticides are:

- Herbicides:
  - Triazines.
  - Carbamates.
  - Amides.
  - Phenoxyalkyl acids.
  - Aliphatic acids.
- Insecticides:
  - Organophosphates.
  - Chlorinated hydrocarbons.
  - Arsenic-containing compounds.
  - Pyrethrum.
- Fungicides:
  - Mercury-containing compounds.
  - Thiocarbamates.
  - Copper sulfate.

These chemicals pose several health risks to humans. Examples of health hazards related to pesticides include diseases of the central nervous system, immune system diseases, cancer, and birth defects.

What are the Processes that Cause Soil Pollution?

Soil pollution can be broadly classified into two categories:

- Naturally caused soil pollution.
- Anthropogenic soil pollution (caused by human activity).

Natural Pollution of Soil

In some extremely rare processes, some pollutants are naturally accumulated in soils. This can occur due to the differential deposition of soil by the atmosphere. Another manner in which this type of soil pollution can occur is via the transportation of soil pollutants with precipitation water.

An example of natural soil pollution is the accumulation of compounds containing the perchlorate anion (ClO_4^-) in some dry, arid ecosystems. It is important to note that some contaminants can be naturally produced in the soil under the effect of certain environmental conditions. For example, perchlorates can be formed in soils containing chlorine and certain metals during a thunderstorm.

Anthropogenic Soil Pollution

Almost all cases of soil pollution are anthropogenic in nature. A variety of human activities can lead to the contamination of soil. Some such processes are:

- The demolition of old buildings can involve the contamination of nearby soil with asbestos.
- Usage of lead-based paint during construction activities can also pollute the soil with hazardous concentrations of lead.
- Spillage of petrol and diesel during transportation can contaminate soils with the hydrocarbons found in petroleum.
- Activities associated with metal casting factories (foundries) often cause the dispersion of metallic contaminants into the nearby soils.

- Underground mining activities can cause the contamination of land with heavy metals.
- Improper disposal of highly toxic industrial/chemical waste can severely pollute the soil. For example, the storage of toxic wastes in landfills can result in the seepage of the waste into the soil. This waste can go on to pollute groundwater as well.
- Chemical pesticides contain several hazardous substances. Excessive and inefficient use of chemical pesticides can result in severe soil pollution.
- Sewage produced in urbanized areas can also contaminate soil (if not disposed of correctly). These wastes may also contain several carcinogenic substances.

Other forms of waste that can pollute soil include nuclear waste, e-waste, and coal ash.

What are the Negative Consequences of Soil Pollution?

Soil pollution harbours a broad spectrum of negative consequences that affect plants, animals, humans, and the ecosystem as a whole. Since children are more susceptible to diseases, polluted soil poses a greater threat to them. Some important effects of soil pollution are discussed below.

Effects on Human Beings

Soil contaminants can exist in all three phases (solid, liquid, and gaseous). Therefore, these contaminants can find their way into the human body via several channels such as direct contact with the skin or through the inhalation of contaminated soil dust The short term effects of human exposure to polluted soil include:

- Headaches, nausea, and vomiting.
- Coughing, pain in the chest, and wheezing.
- Irritation of the skin and the eyes.
- Fatigue and weakness.

A variety of long-term ailments have been linked to soil pollution. Some such diseases are:

- Exposure to high levels of lead can result in permanent damage to the nervous system. Children are particularly vulnerable to lead.
- Depression of the CNS (Central Nervous System).
- Damage to vital organs such as the kidney and the liver.
- Higher risk of developing cancer.

It can be noted that many soil pollutants such as petroleum hydrocarbons and industrial solvents have been linked to congenital disorders in humans. Thus, soil pollution can have several negative effects on human health.

Effects on Plants and Animals

Since soil pollution is often accompanied by a decrease in the availability of nutrients, plant life ceases to thrive in such soils. Soils contaminated with inorganic aluminium can prove toxic to plants. Also, this type of pollution often increases the salinity of the soil, making it inhospitable for the growth of plant life.

Plants that are grown in polluted soil may accumulate high concentrations of soil pollutants through a process known as bioaccumulation. When these plants are consumed by herbivores, all the accumulated pollutants are passed up the food chain. This can result in the loss/extinction of many desirable animal species. Also, these pollutants can eventually make their way to the top of the food chain and manifest as diseases in human beings.

Effects on the Ecosystem

- Since the volatile contaminants in the soil can be carried away into the atmosphere by winds or can seep into underground water reserves, soil pollution can be a direct contributor to air and water pollution.
- It can also contribute towards acid rain (by releasing huge quantities of ammonia into the atmosphere).
- Acidic soils are inhospitable to several microorganisms that improve soil texture and help in the decomposition of organic matter. Thus, the negative effects of soil pollution also impact soil quality and texture.
- Crop yield is greatly affected by this form of pollution. In China, over 12 million tons of grain (worth approximately 2.6 billion USD) is found to be unfit for human consumption due to contamination with heavy metals.

References

- Earths-layers-crust-mantle-core: pmfias.com, Retrieved 11, March 2020
- Soil-Forming-Processes-314500073: researchgate.net, Retrieved 06, June 2020
- Soil-profile, biology: byjus.com, Retrieved 26, April 2020
- Soil-pollution, chemistry: byjus.com, Retrieved 10, January 2020

3

Hydrosphere and its Chemistry

The total amount of water on a planet is termed as the hydrosphere. It includes the water present on the surface, in the air and under the ground. A few areas of focus within this field are water cycle and water pollution. The topics elaborated in this chapter will help in gaining a better perspective about these aspects of hydrosphere and its chemistry.

Water Cycle

The water cycle is the continuous movement of the water all around the Earth. The water cycle is like a big circle and doesn't really have a starting point. There are 4 main stages involved in the water cycle, i.e., evaporation, condensation, precipitation and runoff. So, how does this cycle works? When the sun shines, the water from the ocean or lake evaporates due to heat from the sun.

When water evaporates, it turns into water vapor and goes up into the atmosphere. This water vapor gets together with other water vapor and turns into a cloud. When clouds get dense, they drop the water back to Earth in some form of precipitation like rain, snow, hail or sleet. When the water falls back down to the Earth, they find their way on the ground surface into puddles, streams and rivers. Again this water will evaporate, and the whole cycle will start again.

Water is one of the world's most valuable resources and one that is becoming increasingly difficult to find in drinkable quality. All water in the world is subject to what is known as the water cycle, or the hydrologic cycle, or the H_2O cycle, which is the process by which water moves around the world.

Different Stages of the Water Cycle

Water goes through three different stages in the water cycle. It can be a liquid (water), a gas (water vapor) or a solid (ice). These three states are interchangeable, as water can freeze into ice or evaporate into water vapor, water vapor can condense as water, and ice can melt into water. The water cycle consists of a number of steps that see the water go through each of these states.

Evaporation

Water is found in lakes, oceans, swamps, and soil, as well as in all living creatures and plants. When heat is applied from the sun, through exertion, or by artificial means, the water molecules become excited and spread out. The loss of density is called 'evaporation,' and it sees the water rise into the air forming clouds of water vapor.

Normally, the evaporation of water occurs when the water hits the boiling point, around one hundred degrees centigrade.

However, in places where the air pressure and humidity are lower, far less heat energy is needed to evaporate the water because there is less pressure holding the water molecules together. The water that evaporates from the oceans is not salty, as the salt is too dense and heavy to rise with the water vapor, which is why water from rivers and lakes is not salty.

Snow and ice can actually turn into water vapor without first turning into water. This process is called 'sublimation,' and it results from low humidity and dry winds. This usually occurs at the peaks of mountains or other high-up places, as the lower air pressure means that less energy is needed to sublimate the ice into water vapor.

Some of the highest peaks on the earth, such as Mount Everest, have all of the necessary components for sublimation, namely: strong sunlight, low temperatures, low air pressure, strong wind, and low humidity. If you've ever seen dry ice, which has a fog pouting off of it, this is an example of sublimation in action.

When water evaporates off of a plant's leaves, the process is known as 'evapotranspiration.' A large percentage of the water in the atmosphere is produced by this process due to the large areas covered by plants and trees across the planet. While about ninety percent of the water vapor in the world comes from lakes, oceans, and streams, the remaining ten percent is comprised of the various plant life around the world.

Condensation

The water vapor that has risen into the sky cools significantly when it comes into contact with the cooler air found up high. The vapor becomes a cloud, which is pushed around the world by moving air currents and winds.

If the water vapor cools to anything above zero degrees centigrade, it will condense as water. Essentially, the water vapor will start to condense on the surface of tiny particles of dust and dirt that rose with the vapor during the process of evaporation. These tiny droplets will start to fall into one another and merge, producing a larger droplet.

When a droplet is large enough, gravity will pull it down at a rate that exceeds the updraft in the cloud, leading to the droplet falling out of the cloud and onto the ground below. This process is called 'precipitation', or – more commonly – rainfall.

If precipitation occurs in conditions that are particularly cold or have very low air pressure, then these water droplets can quite often crystallize and freeze. This causes the water to fall as solid ice, known as hail, or as snow. If the conditions are in between those associated with snow and rain, the droplets will fall as icy cold, half-frozen water commonly referred to as sleet.

Precipitation

The water that has fallen as rain is absorbed into the ground through a process known as 'infiltration.' Soil and other porous materials can absorb great deals of water this way, while rocks and other harder substances will only retain a small amount of water.

When the water infiltrates the soil, it will move in all directions until it either seeps into nearby streams or else sinks deeper into what is known as 'groundwater storage.' This is where the water that does not seep out or evaporate joins up under the ground, saturating the smallest nooks and crannies of rock and soil under the ground. These formations are also known as 'aquifers' and explain why sometimes the ground underneath the topsoil is damp or sodden.

When an aquifer becomes too full, it starts to leak out onto the surface, forming what is known commonly as a 'spring.' These can often be found in formations of porous or brittle rock, which can crack following slightly acidic rainfall. Should the water be located near a volcano or any source of natural thermal energy, it will form a hot spring.

Runoff

After the water has fallen and the soil has become saturated, or the snow has melted, the water follows gravity and falls down any hills, mountains, or other inclines to form or join rivers. This process is known as 'runoff,' and it is how water comes to rest in lakes and returns to the ocean. The waterfalls according to the incline of the place from which it is falling, and when several threads of water meet, they form a stream.

The direction in which the water moves is known as 'streamflow', and it is central to the concept of the currents within rivers and streams. These streams and rivers will run off eventually to either form lakes or rejoin the ocean, depending on their proximity to the ocean.

Due to the amount of water stored in snow or ice, a sudden increase in the heat can lead to flooding due to the water suddenly melting and running off at an alarming rate. This is why flooding can occur so easily during a warm spring following a particularly cold and biting winter.

When more snow falls than evaporates or sublimates, the ice will compact densely to form what is known as 'ice caps.' The ice caps and glaciers located in the coldest regions of the world are the biggest collections of ice in the world, and are slowly starting to shrink as the water in which they sit is becoming warmer.

This happens in a cyclical manner, with no beginning or end. As precipitation happens in one part of the world, evapotranspiration is happening somewhere else. The cycle of water never ends, which is why the oceans stay level, there are always clouds somewhere in the sky, and drinkable water doesn't suddenly just run out.

At some point, any water that has infiltrated will be released, any that has joined the ocean will evaporate, and even a large portion of the ice caps may melt, releasing water that has been stored for millions of years back into the water cycle with potentially cataclysmic effects.

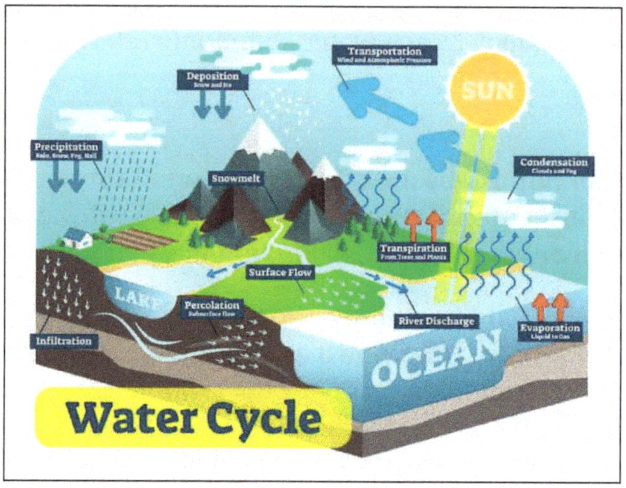

The Importance of the Water Cycle

We cannot Live without Freshwater

All life is dependent upon water. Water makes up 60 to 70 percent of all living matter, and humans cannot live without drinking water for more than a week. The water cycle is an extremely important process because it enables the availability of water for all living organisms and regulates weather patterns on our planet. If water didn't naturally recycle itself and distribute freshwater all over the earth's surface, we would run out of clean water, which is essential to life.

Freshwater seems abundant, but when accounting for all the water on Earth, it's in limited supply. Just three percent of the water on our planet is freshwater. A majority of this water, about two percent of the world total, is contained in glaciers and ice sheets or stored below ground. The remaining one percent is found in lakes, rivers and wetland areas or transported through the atmosphere in the form of water vapor, clouds and precipitation.

Purification

Evaporation and infiltration benefit human, animal and plant life by purifying water. When water evaporates, the pollutants and sediments in it are left behind. As water undergoes infiltration, the ground purifies it of pollutants and contaminants. Glaciers, ice, and snow can act as stores of freshwater for both humans and other organisms. Even aquatic life needs water purified, as saltwater must be within certain pH and saline ranges.

Distribution

Perhaps most important, the water cycle distributes water all over the earth's surface, though unevenly. This is important because if water was not distributed, gravity would push it all into the lowest place, the oceans. The water cycle continually feeds freshwater to all life on the planet: humans, animals and plants. Runoff contributes to rivers, other freshwater bodies, and eventually, the ocean, sustaining freshwater and marine life.

Water Cycle Sustains Life

Plants wouldn't grow without precipitation and therefore anything consuming the plants wouldn't survive and so forth. The water cycle process sustains life and creates the ecosystems around us. Some organisms are very sensitive to changes in the water cycle. A prolonged drought can destroy a population of plants or a certain salamander species may require a set amount of soil saturation in order to avoid desiccation.

The Water Cycle Drives Biogeochemical Cycle

The water cycle patterns and rainfall have major effects on Earth's ecosystems. Rainfall and surface runoff play important roles in the cycling of various elements, also known as the biogeochemical cycle.

In ecology and Earth science, a biogeochemical cycle or substance turnover or cycling of substances is a pathway by which a chemical substance moves through biotic (biosphere) and abiotic (lithosphere, atmosphere, and hydrosphere) compartments of Earth.

These include calcium, carbon, hydrogen, mercury, nitrogen, oxygen, phosphorus, selenium, and sulfur, molecular cycles for water and silica. In particular, surface runoff helps move elements from terrestrial, land-based, to aquatic ecosystems.

Effects on Climate

The water cycle is powered from solar energy, and 86% of the global evaporation occurs from the oceans, reducing their temperature by evaporative cooling. Without the cooling, the effect of evaporation on the greenhouse effect would lead to a much higher surface temperature of 67 °C (153 °F) and a warmer planet.

Properties of Water

Water has chemical formula H_2O. Water is composed of one atom of oxygen with two atoms of hydrogen. Water is a polar molecule with oxygen having a partial negative charge and each hydrogen atom having partial positive charge. Water is a tasteless, odorless liquid and appears colorless in small quantities. The composition of seawater, river water, lake water and groundwater is given in below tables. Seawater contains mainly Na^+ and Cl^- and river water contains mainly Ca^{2+} and HCO_3^-.

Table: Composition of seawater and river water (mmol kg-1).

Element	Seawater (mmol kg-1)	River water (mmol kg-1)
Na	468.0	0.26
Mg	53.1	0.17
Ca	10.3	0.38

K	10.2	0.07
Sr	0.09	----
Cl	546.0	0.22
SO_4	28.2	0.11
HCO_3	2.39	0.96
Br	0.84	---

Table: pH and concentration ranges of the ions (mg L^{-1}) in ground water.

Composition of ground water	
Parameters / Ions	Ranges
Ca^{+2}	200 - 240
Mg^{+2}	58.3 - 87.5
Na^{+}	28 - 35
K^{+}	3.7 - 4.7
HCO_3^-	150 - 298
Cl^-	188.9 - 212
F^-	< 0.5
NO_3^-	330 - 400
SO_4^{2-}	103 - 202
Cu	0.01 - 0.04
Pb	< 0.1
Zn	0.60 - 1.30
Fe	0.33 - 1.12
Mn	0.01 - 0.03
pH	7.36 - 7.80

Table: Composition of Lake water (mg L^{-1}).

Composition of Lake water	
Calcium	241
Magnesium	7200
Sodium	836,00
Potassium	4070
Bicarbonate	251
Sulfate	16,400
Chloride	140,000
Silica	48
TDS	254,000
pH	7.4

Unique Properties of Water

It is the only abundant common pure compound, which is a liquid. It has many unusual chemical and physical properties. The Earth appears to be a majestic blue marble from top. This is due to the abundance of water on the earth surface. While water itself is not blue, water gives off blue light upon reflection.

In simplest terms, water makes up about 71% of the Earth's surface, while the other 29% consists of continents and islands. To break the numbers down, 96.5% of all the Earth's water is contained within the oceans as salt water, while the remaining 3.5% is freshwater lakes and frozen water locked up in glaciers and the polar ice caps. Of that fresh water, almost all of it takes the form of ice: 69% of it, to be exact. If you could melt all that ice, and the Earth's surface was perfectly smooth, the sea levels would rise to an altitude of 2.7 km. Water is the clear sparkling fluid which covers three quarters of the earth's surface. Not to mention water is better known as the basis of life. Water is the possessor of the world's most recognizable chemical formula (H_2O).

Water molecules are attracted to each other, and form hydrogen bonds. These weak interactions determine the most physical property of water and many of its chemical properties too.

A large part of the mass of most organisms is simply water. In human tissues, the percentage of water ranges from 20% in bones to 85% in brain cells. The water content is greater in embryonic and young cells and decreases on ageing. About 70% of our total body weight is water.

Water: The Universal Solvent

Water has the ability to dissolve most of the substances; some have high solubility whereas some are only sparingly soluble. It is, therefore, called a universal solvent. In rainwater, the substances and atmospheric gases present in air get dissolved.

The important physical properties of water are in table below.

Property	Value
Molar Mass	18.015
Molar Volume	55.5 moles/liter
Boiling Point (BP)	100°C at 1 atm
Freezing Point (FP)	0°C at 1 atm
Triple Point	273.16 K at 4.6 torr
Surface Tension	73 Dynes at 20°C
Vapor Pressure	0.0212 atm at 20°C
H of Vaporization	40.63 kJ/Mol
H of Fusion	6.013 kJ/Mol
Heat Capacity (CP)	4.22 kJ/kg.K
Dielectric Constant	78.54 at 25°C
Viscosity	1.002 centipoise at 20°C
Density	1 g/cc
Density Maxima	4°C

Global Water Distribution

For an estimated explanation of where Earth's water exists, look at the chart below. By now, you know that the water cycle describes the movement of Earth's water, so realize that the chart and table below represent the presence of Earth's water at a single point in time. If you check back in a thousand or million years, no doubt these numbers will be different.

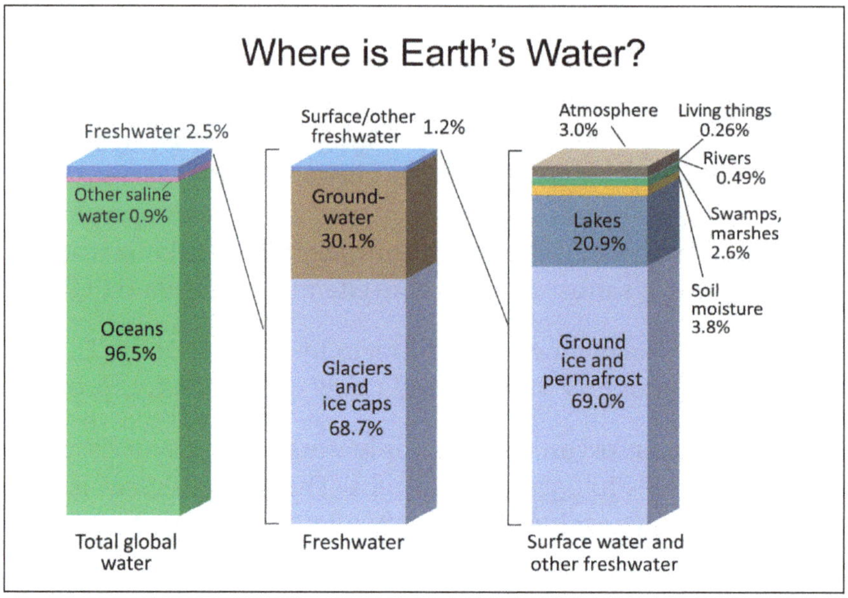

Notice how of the world's total water supply of about 333 million cubic miles (1,386 million cubic kilometers) of water, over 96 percent is saline. And, of the total freshwater, over 68 percent is locked up in ice and glaciers. Another 30 percent of freshwater is in the ground. Thus, rivers and lakes that supply surface water for human uses only constitute about 22,300 cubic miles (93,100 cubic kilometers), which is about 0.007 percent of total water, yet rivers are the source of most of the water people use.

Table: One estimate of global water distribution(Percents are rounded, so will not add to 100).

Water source	Water volume, in cubic miles	Water volume, in cubic kilometers	Percent of freshwater	Percent of total water
Oceans, Seas, & Bays	321,000,000	1,338,000,000	--	96.54
Ice caps, Glaciers, & Permanent Snow	5,773,000	24,064,000	68.7	1.74
Groundwater	5,614,000	23,400,000	--	1.69
Fresh	2,526,000	10,530,000	30.1	0.76
Saline	3,088,000	12,870,000	--	0.93
Soil Moisture	3,959	16,500	0.05	0.001
Ground Ice & Permafrost	71,970	300,000	0.86	0.022
Lakes	42,320	176,400	--	0.013
Fresh	21,830	91,000	0.26	0.007
Saline	20,490	85,400	--	0.006
Atmosphere	3,095	12,900	0.04	0.001
Swamp Water	2,752	11,470	0.03	0.0008
Rivers	509	2,120	0.006	0.0002
Biological Water	269	1,120	0.003	0.0001

Water Pollution

Inorganic Pollutants

Some of the heavy metals are amongst the most harmful of the elemental pollutants and are of particular concern as they are dangerous to humans. These elements are generally the transition metals and few of the representative elements in the lower right-hand corner of the periodic table such as lead and tin. Few essential elements come under the category of heavy metals like iron as well as toxic metals like cadmium and mercury. Most of these form bonds with sulfur groups in enzymes resulting in disruption enzyme function. Protein carboxylic acid ($-CO_2H$) and amino ($-NH_2$) groups are also chemically bound by heavy metals. Cadmium, copper, lead, and mercury ions bind to cell membranes, disturbing transport processes through the cell wall.

Cadmium

The solubility of cadmium compounds in water is affected by the acidity of the water and the nature of the source. Cadmium enters water from industrial discharges and mining wastes. Cadmium is very similar to zinc and undergoes geochemical processes along with zinc. It is present in the +2 oxidation state. The levels of cadmium could be higher in areas supplied with soft water of low pH. Acute effects have been seen where food has been contaminated by cadmium. Acute cadmium poisoning in humans has very serious effects. Severe gastrointestinal upsets; bronchitis, emphysema, anemia, high blood pressure and kidney damage have been reported. Generally cadmium replaces zinc in certain enzymes thereby altering its structure and impairing its catalytic activity. Itai-itai disease, the mass cadmium poisoning in Toyama Prefecture, Japan, started around 1912. This cadmium poisoning caused kidney failure softening of bones. The disease caused severe pains in the joints and spine. The cadmium was discharged into rivers by mining companies in the mountains. The mining companies were later sued for the damage. Itai-Itai disease is one of the biggest pollution diseases of Japan.

Lead

In aquatic systems, lead has been found to be quite toxic to many organisms even at small concentrations. Acute lead poisoning in humans causes severe dysfunction in the kidneys, the brain, liver, reproductive system and central nervous system and causes sickness or death. Nowadays lead is not a serious pollutant in drinking water except in those cases where lead pipes are still being used.

Mercury

This metal generates a lot of concern as a pollutant. Mercury is discharged into natural waters from industrial effluents. Metallic mercury is used as an electrode in the manufacture of caustic soda and chlorine and in laboratory vacuum apparatus and thermometers. Industries like batteries, vapour lamp fertilizers, pesticides etc. also use mercury compounds. Mercury has a great tendency to bio accumulate in the forms of methyl mercury and shows a greater capability of bio magnification through food chains.

Because of their resistance to degradation and their mobility, the alkyl mercury compounds are generally considered to be more of an environmental threat than either the aryl or inorganic compounds. The toxicity of mercury was tragically illustrated in the Minamata Bay area of Japan during the period 1953-1960. A total of 111 cases of mercury poisoning and 43 deaths were reported among people who had consumed seafood from the bay that had been contaminated with mercury waste from a chemical plant that drained into Minamata Bay.

Metalloids

The most significant water pollutant in this category is arsenic. It is a naturally occurring element found in the earth's crust and within many ores. It is released to the water bodies through industrial smelting and natural processes like soil erosion, weathering and mineral leaching. The toxicity of arsenic compounds is well known since ancient times. It occurs as organic and inorganic compounds in drinking water and the inorganic forms are more toxic than the organic forms. The toxicity also depends on the oxidation state and As (III) compounds (arsenites) are ten times more toxic than As(V) compounds (arsenates). It gets accumulated in the tissues and causes gastroenteritis, skin lesions, and skin cancer and multi system organ failure. The toxicity is mainly due to the deactivation of enzymes containing – SH groups as arsenic binds to this group. It also adversely affects generation of ATP and coagulates proteins. Water pollution due to arsenic is noted in West Bengal, some north eastern states of India and in our neighboring country Bangladesh. The situation in villages of Bangladesh has been classified as a 'real disaster' and the arsenic levels in tube well water range from 0.05 mg/L to 10mg/L which is well above the WHO maximum permissible limit of 50μg/L. According to WHO and UNICEF arsenic has been labeled as a 'ticking time bomb'. The diabolic situation of the two nations is a result of years of negligence. While setting up tube wells in these two nations, the responsible organizations had not paid enough attention towards the quality check of underground water and did not even conduct tests for arsenic.

The toxic effects of some pollutants are summarized in table below.

Substances	Effects
Arsenic	Very poisonous, can cause cramps, paralysis and even death.
Ammonia, ammonium hydroxide and ammonium salts	Toxicity increases with increase in pH; toxicity to fish increased 200% when pH was raised from 7.4 to 8.0
Cadmium	Synergistic with copper, the toxicity of zinc and cadmium is additive.
Calcium	Antagonistic to lead, zinc, aluminium and toxic solutions of sodium, potassium and magnesium.
Chlorine	Toxicity increases with decrease in pH, combines with thiocyanate to form a non-toxic effluent to toxic form, paraoxon, combines with phenols and ammonia to from a chlorophenols and chloramines, both toxic substances.
Fluorides	Toxicity increases with decreases in pH. Above 1 mg/L causes fluorosis, molting of teeth enamel, nervous and skeleton disorders.
Copper	A high pH above 7 may reduce the toxicity by forming a precipitate of copper hydroxide; copper precipitates are also formed in water with high hardness; low oxygen increases the toxicity of copper, synergistic with chlorine, zinc, cadmium and mercury, antagonistic towards sodium nitrite and decreases the toxicity of cyanide.

Cyanide, hydrogen cyanide	Below pH 8, cyanide is in un-dissociated state and is toxic; toxicity increases 2-3 times with 10°C rise in temperature; toxicity increases with fall in oxygen, highly toxic with zinc or aluminium.
Hydrochloric acid	Toxicity only when pH is below 5.
Manganese	Nickel and manganese are antagonistic.
Mercuric chloride	Synergistic with small amounts of sodium chloride; however, the effect is antagonistic with high concentration of sodium chloride.
Lead	Synergistic with calcium and a cumulative poison, causes loss of appetite, constipation. Abdominal pain, nervous disorder and brain damages.
Sodium	Antagonistic towards the toxicities of potassium, ammonium and calcium salts.
Sodium ferrocynaide	Decompose under sunlight to produce cyanide and hydrogen cyanide which are more toxic.
Sodium sulphide	Toxicity increases with decrease in pH from 9 to 6.
Sulphuric acid	Toxic when pH is below 5.
Zinc	Toxicity increases with increasing temperature, and decreasing oxygen, toxicity decreases with increasing hardness, synergistic with nickel cyanide and copper, toxicity of zinc and cadmium is additive.
Iron and manganese	Stain fabrics cause off tastes and odor, modify colors in dyeing. Manganese causes paralysis of lower limbs.

Acid Mine Drainage

This is a very common and damaging problem in aquatic environment and is similar to acid rain. Coal mines discharge substantial quantities of sulphuric acid and iron (III) hydroxide into local water bodies through seepage. On exposed coal mine surfaces, minerals containing sulphur (most notably iron pyrites, FeS_2) come into contact with air and water, forming sulphuric acid that has been carried into streams by waters from the mines. FeS2 is stable in absence of air but in presence of air during mining operations the oxidation reactions take place and produce large amount of sulphuric acid.

$$2FeS_2 + 2H_2O + 7O_2 \rightarrow 2Fe^{2+} + 4HSO_4^-$$

$$4Fe^{2+} + O_2 + 4H^+ \rightarrow 4Fe^{+3} + 2H_2O$$

The roles of microorganisms are important in the overall process. The rate of these reactions is slow at a pH below 3.5 but in presence of thiobacillus ferro-oxidants bacteria the rate of the reaction increases. In the pH range of 3.5 to 4.5, the reaction may be catalyzed by a variety of bacteria called metallogenium. Ferric ions produced in the above reaction dissolves the pyrites further and together with the second reaction constitute a cycle for the dissolution of pyrite.

$$FeS_2 + 14Fe^{3+} + 8H_2O \rightarrow 15Fe^{2+} + 2SO_4^{2-} + 16H^+$$

$$Fe^{+3} + 3H_2O \leftrightarrow Fe(OH)_3 + 3H$$

Acid mine drainage has been the primary source of pollutants that increase the acidity of water bodies. This adversely affects thousands miles of streams and has been one of the most significant causes of water degradation. The acidic water is toxic to the plants and animals of an aquatic ecosystem. It also damages biological productivity. Alkalies discharged by industries such as textiles, tanneries and coke oven operations can destroy aquatic life.

Oil Pollution

Oceans are being polluted on daily basis from oil spills, routine-shipping, run offs and dumping. 12% of the oil pollution in oceans is caused by Oil spills and rest is from drains, dumping and shipping travel. Oil is annually consumed in large quantity therefore its use increases day by day. The total annual influx of petroleum hydrocarbons into ocean is about 10 million metric tonnes. The major amount of these petroleum (oil) products released into water bodies are due to transportation related activities. Oil tanker accidents are responsible for oil spills. They cause severe problem because high concentration of oil spill during such accidents. It causes acute localized problems but can be hazardous for marine wildlife in the local area such as birds, fish and other aquatic species. Oil and water are immiscible therefore iyt forms a thick layer at the surface of water in oceans and river. This causes suffocation to fish and also creates problems for marine birds by making them stop from flying. The layer of oil at the surface of water of oceans also causes blockage of sunlight for aquatic plants and they cannot undergo the process of photosynthesis. Heavy oil residues that are not degraded get deposited on sediments and end up as tar lumps or tar balls. The marine organisms experience direct lethal toxicity and disturb the physiological and behavioral activities. Birds are especially vulnerable to damage from oil coating; they lose insulation as the insulative compounds dissolve in oil and the birds freeze to death in winters.

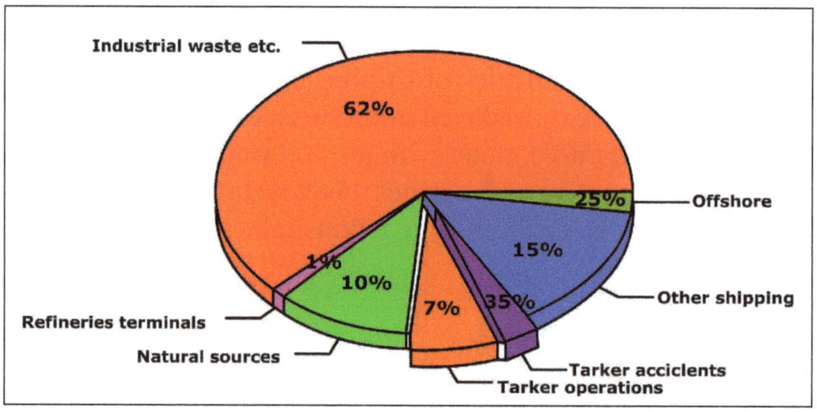

Source of oil pollution into the sea.

Thermal Water Pollution

Thermal pollution is defined as sudden increase or decrease in temperature of a natural body of water which may be ocean, lake, river or pond by human influence. These sudden changes in temperature adversely affect water quality and aquatic life. The most important anthropogenic source of thermal water pollution is the coal fired power plant and industries which release large quantities of heat directly in the water body. These plants draw water from nearby source to keep machines cool and then release water back to the source with higher temperature. When heated water returns to the river or ocean, the water temperature rises sharply. Waste heat is an inevitable by-product of the power plants. Approximately 60% of the heat is rejected by fossil fuel power plants. On the other hand the nuclear power plants rejects heat approximately upto 70%. This is the reason power plants are preferred to build away from rivers, lakes, or seashores for a ready supply of cooling water. The condenser cooling water should not have temperature more than 5°C higher than the intake water temperature. The maximum temperature of a thermal discharge (t_{ww}) in a water basin can be evaluated by using the formula:

$$t_{ww} = (aQ/q+1)t_p + t_{max}$$

Where,

t_{max} = maximum temperature of water body at summer,

t_p = permissible temperature rise,

Q = flow rate of water above the discharge in the water body,

q = flow rate of thermal discharge,

a = mixing coefficient.

Soil erosion and deforestation are other major factors that cause thermal pollution. Trees and plants prevent sunlight from falling directly on water bodies. When deforestation or soil erosion occur then these water bodies are directly exposed to sun light and thus absorbing more heat and rising its temperature. Natural disaster like volcanoes, lightning and geothermal activity under the seas and oceans can trigger warm lava to raise the temperature of water bodies.

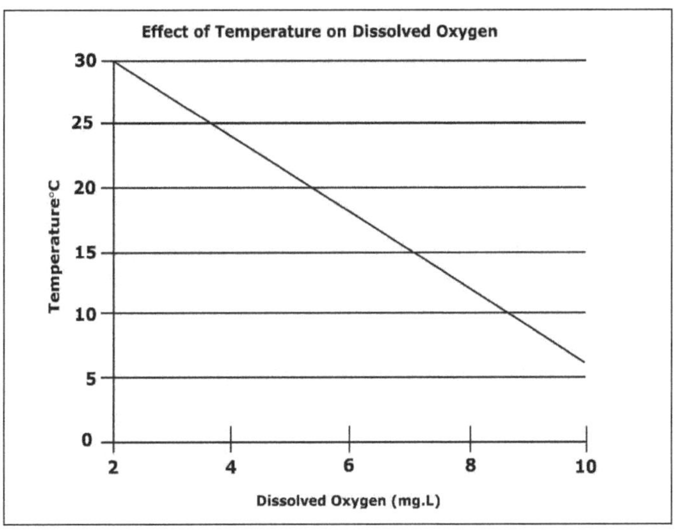

Effects of Thermal Pollution

- The amount of dissolved oxygen in the water decreases. The figure shows the effect of increasing temperature on the dissolved oxygen. Since metabolism requires oxygen, some species may be eliminated entirely if the water temperature rises by 10° C. Additionally, dissolved oxygen is the key to decompose organic wastes by microorganisms. Heating a water body will impair this decomposition process and further increase the water pollution.

- The rate of chemical reactions increases by raising temperature of water body. The life cycles of many aquatic organisms are closely and delicately effected by water temperature. Trout eggs hatch in 165 days when incubated at 20°C. When water temperature has 30°C, only 32 days are needed and no hatching takes place at water temperature in excess of 33°C. Such a result can be disastrous to fish population.
- The survival of aquatic life becomes difficult. False temperature cues have been given to aquatic life. Fish are often disturbed, migrate, and spawn in response to temperature cues. When water temperature is artificially changed, the disruption of aquatic organisms' normal activities and patterns can be catastrophic. There may be large-scale migration to an environment more favorable to their survival. The addition of new species of fish will change the eco-balance of the migrated area. Thermal pollution can also increase the susceptibility of aquatic organisms to parasites, toxins and pathogens, making them vulnerable to various diseases. If these processes are continued for long period of time, huge amount of bacterial growth results in algal bloom and decreases the amount of oxygen of water.
- Lethal temperature limits may be exceeded. Fish are able to acclimatize themselves to moderate temperature changes (below lethal level) if the change has been not a sudden one. Normally a moderate temperature change over a 20 to 40 hour period could be handled by the fish. For example 95% of the eggs of largemouth bass perish when suddenly transferred from water at 35-39 °C into water at 45°C. However, if similar eggs are kept in water and the temperature gradually increased to 45°C over a 30 to 40 hour time period, 80% of the eggs survive.

Radioactive Pollution

Radioactive pollution is caused by the harmful radiation from the radioactive materials like uranium and thorium. The sources of radioactive waste into water bodies are mining and refining of radioactive materials, waste from nuclear power plant, industries and medical waste. Many radioactive wastes are lethal at relatively low concentrations and in minute amount may be mutagenic. A typical uranium ore contains about 2-5 lbs of U3O3 per tonne so that large amounts of ore are processed for extraction of uranium. As a result large quantities of uranium tailings are produced, which pose the problem of radioactive pollution.

Radioactive substances can enter the humans with food and water and get accumulated in blood, thyroid gland, liver, bones and muscular tissues. These radioactive materials cause mutations leading to skin cancer or leukemia and they damage the reproductive system and also reduce the effectiveness of certain enzymes. Radium and thorium are chemically similar to calcium and so get absorbed by the bones when taken into the body. Some waters in the Colorado River Basin have had the concentration of 88Ra226 increase to about double the maximum permissible for human consumption.

Based on the concentration of radioactive elements in water body, the waste waters may be classified into three categories:

- Low level waste: Wastes that are within specified concentrations of radioactivity and may arise from the non-nuclear and nuclear industries and typically consist of everyday items that have become contaminated during use by contact with radioactive materials, and waste from decommissioning operations such as building rubble, soil and metal items such as framework, pipe-work and reinforcement. These wastes may be solid, liquid or gaseous and have very low radioactivity and produced in large volumes.

- High level wastes: These wastes are usually liquid or solid wastes having very high level of radioactivity and released in low volume. These wastes directly cannot be released into the environment without treatment.
- Intermediate level waste: These wastes are also in liquid or solid state with high enough radio activity. These are also released in low volume with proper treatment.

Measurement of Radioactivity

The radioactivity can be measured in terms of radiation emitted or in terms of radiation absorbed by an organism. It is expressed in terms of Curie (Ci), Rad and Rem.

The becquerel (Bq) measures the activity of the radioactive source, meaning the number of atoms which, within a particular time frame, transform and emit radiation.

1 Bq = 1 emission of radiation per second,

1 curie = 3.7×10^{10} disintegration per second.

The gray (Gy) measures the absorbed dose, meaning the energy transferred to the material by ionizing radiation upon encountering it.

1 Gy = 1 joule per kilogram,

1rad = 100 ergs per g.

The sievert (Sv) defines ionizing radiation defect on living material. Contrary to the becquerel, the sievert is a very large unit, and we often use sub-multiples: 1 mSv = 1 millisievert = 0.001 Sv A factor called quality factor is used in human health to measure the relative effectiveness of different radiations to produce some biological effects. Dose in Rem = Dose in Rad x RBE For gamma and beta radiation 1 Rem Dose equals to 1 Rad and for alpha radiation 1 Rem dose will be equal to 20 Rad.

Organic Pollutants

Organic pollution of rivers by wastewater discharge from human activities (cities, farming, and industry) affects humans and ecosystems worldwide through the global sanitation crisis. First, untreated urban sewage contains pathogens that cause a variety of diseases, including diarrhea, globally the leading cause of illness and death. As of 2015, up to 2.4 billion people, primarily in sub-Saharan Africa and southern Asia, lack access to proper sanitation. Second, accumulation of organic pollutants in rivers stimulates microbial growth, leading to oxygen depletion and disturbance of the entire river ecosystem.

The level of organic pollution in a river, commonly expressed by the Biochemical Oxygen Demand (BOD), is the result of two counteracting mechanisms: pollutant loading and natural cleaning. Wastewater discharge from cities and intensive livestock farms constitute the main organic pollutant loads into rivers. With rapid urban population growth expected in the next decades, both sources of organic pollution will increase. Although pollution is introduced at wastewater discharge points along the river, impacts extend to downstream populations and ecosystems, as pollutants are transported through the river network. The extent of downstream impacts depends on self-cleaning capacities of rivers via dilution by natural runoff and natural degradation by micro-organisms, as illustrated for two major basins in Figuer below. Changes in river discharge due to climate change affect river dilution capacities, increasing the risk of river pollution in areas that experience reductions in climate wetness. Increases in water extractions to support a growing global population may further decrease river dilution capacities.

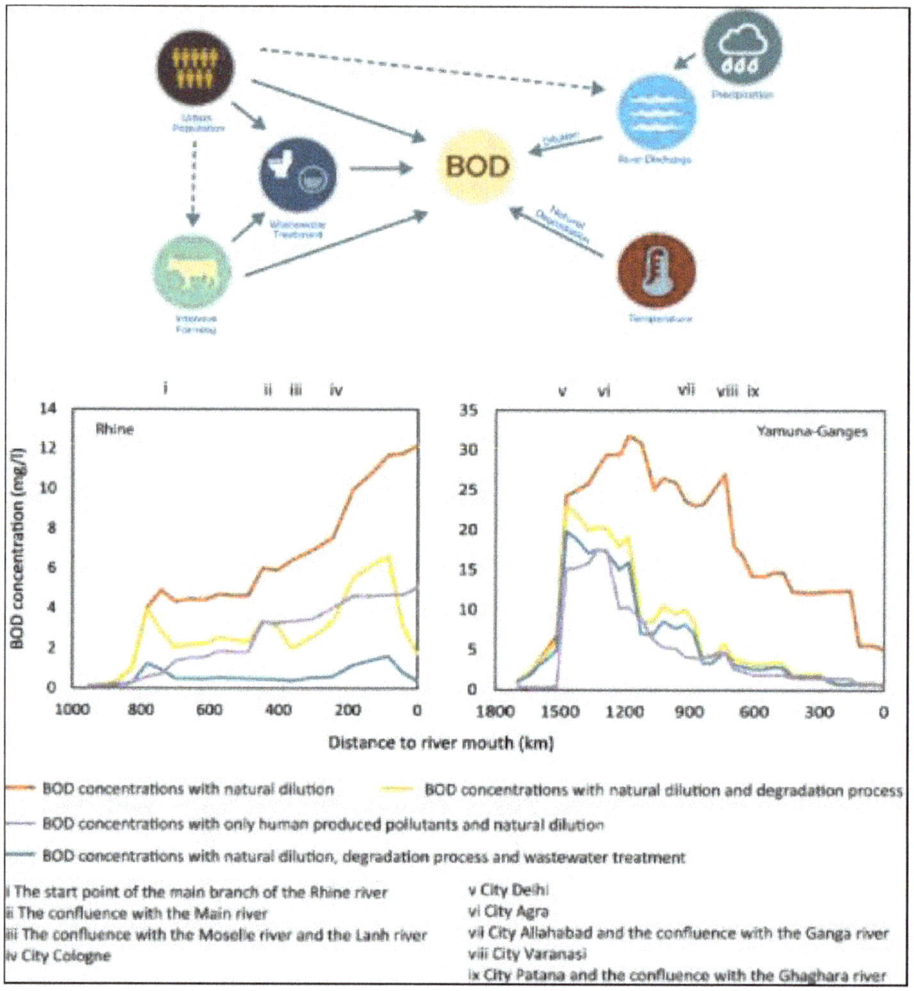

(a) Variables and processes affecting organic pollution of rivers expressed as in-stream Biochemical Oxygen Demand (BOD); (b) Calculated in-stream BOD concentration profiles from headwater to river mouth for the year 2000 along the main stem of the Rhine and Yamuna-Ganges rivers. Without natural degradation and wastewater treatment (purple profiles), BOD concentrations gradually increase along the densely populated Rhine, whereas for the Yamuna-Ganges river a rapid increase near the cities of Delhi and Agra is followed by dilution with freshwater from several large tributaries. Pollutant loading from livestock farming adds significant pollution in both basins (red profiles). When natural degradation is taken into account (yellow profiles), BOD concentrations decrease significantly in both rivers, illustrating the self-cleaning capacity of natural rivers. Wastewater treatment (blue profiles) further reduces BOD concentrations, especially in the Rhine, but also near Delhi and Agra, which have higher rates of wastewater treatment than smaller cities further downstream.

Chemical-based products are found in plastic casing for computers, in athletic-hall flooring, in rubber soles for jogging shoes – the applications are endless. In consequence, a very wide range of chemicals is used in industry today. According to the Organisation for Economic Co-operation and Development (OECD), approximately 100,000 different chemical substances are currently on the market worldwide. In Europe alone, approximately 10,000 chemicals are produced and marketed annually in amounts of greater than 10 tonnes. It is estimated that between 1 and 3 per cent of these chemicals are problematical. These environmentally relevant pollutants include heavy metals such as lead and mercury, which are released into the environment by the burning of oil, mining activities, and industrial emissions and effluents. Persistent organic pollutants, known as POPs, are another problematical substance category.

Fig: During the phocine distemper virus (PDV) epidemics in 1988 and 2002, thousands of dead seals were washed up on German beaches and had to be collected and destroyed.

Poisonous and Persistent – POPs

The Stockholm Convention on Persistent Organic Pollutants adopted in 2001 deals with persistent organic pollutants (POPs), i.e. substances that possess toxic properties and resist degradation. They include pesticides such as DDT and lindane, industrial chemicals such as polychlorinated biphenyls (PCBs), and substances such as dioxins, which are the unwanted by-products of manufacturing and combustion processes. As these substances are highly stable and therefore non-degradable to a large extent, they can be transported over long distances and accumulate in the environment.

POPs cause problems because they are stored in the fatty tissue or organs of animals, where they can have toxic effects. For example, they can disrupt the endocrine system, cause cancer or genetic defects, and weaken the immune system.

Various effects of POPs on marine mammals have been investigated. Studies of Baltic ringed seals (Phoca hispida) and grey seals (Halichoerus grypus) found uterine occlusions, stenoses and tumours, resulting in reduced reproductive ability. Other observed effects included colonic ulcers, as well as reduced bone density, which led to changes in the skeletal system. In seals and porpoises, researchers found indications that POPs depress the immune and endocrine systems.

Humans mainly ingest POPs from food and drinking water, but also from the air (mainly by breathing in dust particles) and through the skin (through direct contact with the chemicals). The highest concentrations of POPs are generally found in marine mammals and humans, both of which are at the top of the food chain.

Fig: Bioaccumulation of toxins in the marine food chain has long been recognized as a problem. The process illustrated here relates to polychlorinated biphenyls (PCBs), a typical environ-mental toxin.

Polyfluorinated Compounds: A Fresh Cause for Concern

Besides the "classic" POPs new types of persistent toxic compounds of non-natural origin were identified in the environment at the end of the 1990s, which could not be detected before as the appropriate technology and analytical methods had not yet been developed. These include polyfluorinated compounds (PFCs), which have been used in a wide variety of every-day applications for more than 50 years. PFCs are mainly used as fluoropolymers in the textile industry, for example, in the manufacture of breathable membranes for outdoor clothing, and in the paper industry in the production of water-, stain- and grease-proof paper (e.g. fast-food packaging). They are also used for surface treatment of furniture, carpets and clothing textiles and in non-stick coatings for cookware (such as Teflon frying pans).

It is believed that a total of six manufacturers have produced around 4500 tonnes of PFCs every year over the past decade: a relatively small amount compared with other chemicals. This group of substances is significant nonetheless, due to its environmentally relevant properties, as some PFCs are highly bio-accumulative in organisms.

At present, more than 350 different PFCs are known to exist. The best-known is perfluorooctanesulfonic acid, more commonly known as PFOS. Based on animal experiments with PFOS, researchers conclude that repeated exposure can have an extremely adverse effect on human health; among other possible effects, it may cause damage to the liver. PFOS may also be carcinogenic, and it is also thought to impair the development of progeny. PFOS therefore recently became the first PFC to be listed as a persistent organic compound (POP) under the Stockholm Convention, which means that it is now on the list of particularly hazardous chemicals for which a worldwide ban is to be imposed.

Occurrence of Polyfluorinated Compounds

Polyfluorinated compounds (PFCs) have been industrially manufactured for around half a century, but it has only recently been possible to detect their presence in the environment due to new chemical and analytical techniques. Natural origins of these chemicals are not known to exist, and yet today PFCs can be detected in water, soil, air and living organisms worldwide – including humans. High levels of PFCs have been found in numerous foods as well as in human blood and breast milk.

The distribution of PFOS in the environment is particularly well-researched. High concentrations of these substances have been detected in fish, seals and sea birds worldwide and, above all, in Arctic polar bears, which are at the top of the food chain. Researchers from Canada and Denmark have reported a sharp rise in PFOS concentrations in liver samples taken from polar bears in Canada, Alaska and Greenland in recent decades. Compared to other environmentally relevant POPs, such as polychlorinated biphenyls, PFCs are found in much higher concentrations. In Swedish studies of human blood from 1994 to 2000, the mean PFC concentration was 20 to 50 times higher than the concentration of the polychlorinated biphenyls and about 300 to 450 times higher than that of hexachlorobenzene, two "classic" organic pollutants that have been recognised as hazardous for decades.

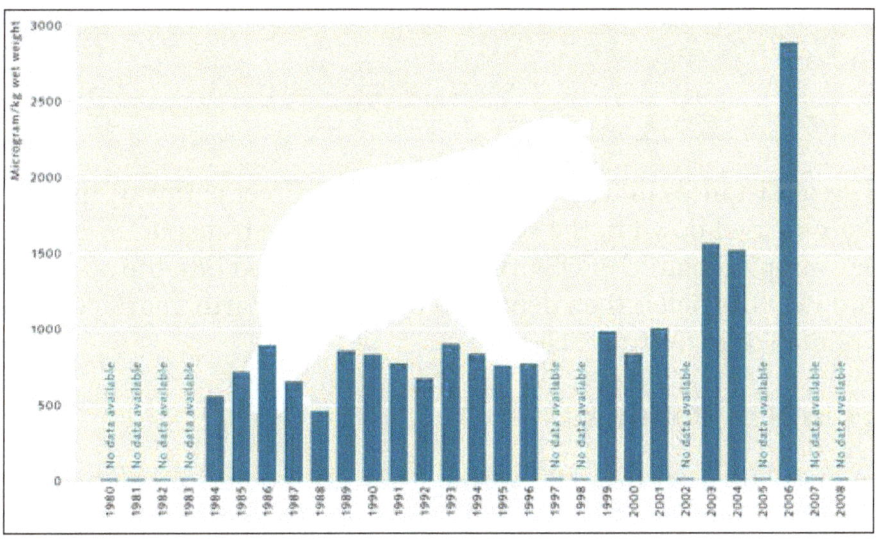

Fig: PFOS concentrations in the livers of East Greenland polar bears have increased significantly in recent years. The measurements were obtained from deep-frozen liver samples.

Transport of PFCs

The detection of PFCs and especially PFOS in marine mammals such as Arctic seals and polar bears, and even in the blood of the Arctic's human inhabitants, the Inuit, raises the question of transportation: How did these substances end up in the sea and even in the Arctic? There are numerous different sources of PFCs. They are released, for example, during the use of the everyday consumer durables – from carpeting, outdoor clothing, cookware and fast-food packaging. However, in Germany, relatively large concentrations of PFCs also enter the rivers from municipal and industrial wastewater treatment plants, which cannot capture these substances. The rivers then wash these substances into the North Sea. From here, they are carried by the main North Sea and Atlantic Ocean currents to the Arctic, where they are ingested by microorganisms in the water and thus enter the food chain, bioaccumulating in larger organisms and finally in the organs of polar bears and humans.

PFCs are also transported long distances through the atmosphere by the movement of air masses. Compounds such as PFOS are not volatile, but volatile precursor compounds escape into the atmosphere during the manufacturing process. Physical and chemical processes that take place in the atmosphere then convert these chemical precursors into stable end products such as PFOS. These are removed from the air by precipitation and enter the seawater in soluble form or bound to dust particles, or are deposited on land or ice surfaces. PFCs can thus travel great distances and can be detected in the environment a long way from their place of origin or use.

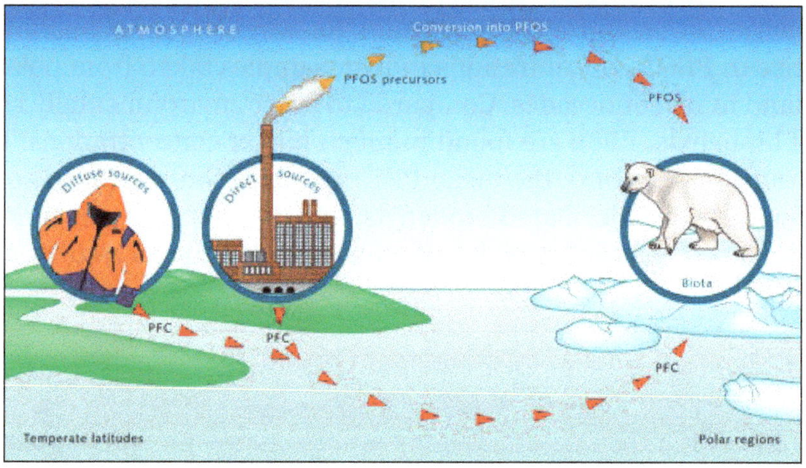

PFCs can travel great distances in water or air. Through a direct pathway, they enter the rivers in wastewater and are carried down to the sea. They can also be transported indirectly through the atmosphere. For example, volatile PFOS precursors are released into the atmosphere, where they are converted into PFOS, which is then deposited back on the Earth's surface at the place of origin or elsewhere in rainfall or in dust particles.

Protection from New Pollutants

Today PFCs are distributed all over the world. They are found in water, in the air, in livi ng organ-isms and even in our own bodies. They are likely to persist for generations. This group of sub-stances clearly shows that it is impossible to predict all the environmental impacts, or the delayed effects, of new chemical substances. In the future too, it is likely that some substances that were initially regarded as harmless, but whose undesirable effects can only be discerned after some time has elapsed, will be detected in the marine environment. Nowadays, however, intensive efforts are being made to limit the further global spread of pollutants. For example, risk assessments are carried out before chemicals are licensed for use, in order to determine to what extent they could constitute a hazard. There are also various voluntary renunciation schemes for producers, as well as relevant legislation. In other words, a start has been made.

Effects of Organic Pollutants

Although microorganisms can degrade organic pollutant load in water through a self-purification process involving using of sufficient oxygen, dilution, sedimentation and sunlight.

Plant

Exposure of aquatic plants to organic pollutants is generally through uptake from roots influenced by their low volatility and through plant leaves by contact from air, often a consequence of agricultural spraying with organochemicals. After uptake by plants, organic pollutants are translocated to different parts of the plants, where toxicity may occur. Transport pathways in higher plants generally involve; short distance transport (intracellular and intercellular) and long distance transport (conducting tissue transport). However, some chemical based on their chemistry (e.g water-hating organic chemicals) are only limited in phloem.

Aquatic plant tolerance to uptake of organic pollutants seems to correlate with the ability to deposit large quantities of pollutant metabolites in the 'bound' residue fraction of plant cell walls compared to the vacuole, where enzymatic and metabolic activities may occur. However, toxicity of organic pollutants may be based on plant part viz root and leave. To the leave cell, toxic effects may include cell ultrastructure, biosynthesis, membrane stability and DNA while to the root cell, toxic effects include inordinate mitotic division. Other effects may be on plant physiological and biochemical responses. Some recent studies, found that the system of defense and growth of Chara vulgaris L., Lemna minor L., H. dubia (Bl.) Backer and Potamogeton perfoliatus L respectively are affected by Linear Alkylbenzene Sulphonate, (LAS). Furthermore, the effects are varied among the different aquatic plant species. Similarly, reported that at concentration of 840 mg/L of ammonium nitrate in water, the growth rate, carbon contents, carbon-nitrogen ratio, photochemical cells and induced reactive oxygen stress (ROS) of Lemna minor L (Duckweed) was reduced, resulting in cell mortality of the aquatic plant. A simple indicator of aquatic plants exposure to organic pollutant is seen by the increased ROS production, leading to plants inability to do its regular ecological function of regulating nutrients in aquatic environment. Information regarding the toxic effects of organic pollutants especially POPs on aquatic plant species or macrophyte is very scarce.

Aquatic Animals

Available oxygen in water is reduced organic pollutants. This affects water organisms by causing reduced fitness or death from asphyxiation. Effects also include increased turbidity (especially by petroleum-related wastes) of the water, which reduces the available light for photosynthetic organisms and potentially leading to its death. It can also settle on the benthic and alters the characteristics. Organic pollutants have been detected in marine organisms, including the green mussel, Perna viridis, barnacles, odontocete species and fish species.

Human Health

Environmental xenobiotic compounds have the potential to induce adverse effects on human health. A common example is hydrophobic contaminant like POPs are known to be a potential endocrine disruptor compounds. Although, impact of organic pollutants to human health is yet to be fully examined, evidences still exists which correlates development and manifestation of some chronic diseases to exposure of certain organic chemicals. Particular, cancer cases has been greatly linked. Other toxic effects could be on ovarian function in women, reproductive disorders in both male and female, female breast cancer, blood poisoning, eyes and skin irritation (by exogenous pollutant e.g. LAS).

Chromatography

Chromatography is a method of separation in which the components to be separated are distributed between two phases, one of these is called a stationary phase and the other a mobile phase which moves on the stationary phase in a definite direction. The components of a mixture redistribute themselves between two phases by a process which may be adsorption, partition, ion exchange or size exclusion. The stationary phase can be a solid or a liquid and the mobile phase a liquid, gas or a supercritical fluid. The stationary liquid is held on an inert solid support or coated on the walls of a capillary. The solid whether it is support or an active material is held in the form of a column, coated over a plate or presented as a paper. The latter two types come under the head of two dimensional chromatography.

Classification

If we look into the definition of chromatography it can be classified on the basis of the following:

- The shape of the solid support.
- The nature of the mobile phase.
- The mechanism responsible for separation.

If we go by the shape of support in chromatography, there will be three dimensional (column) and two dimensional (thin layer and paper) chromatography. In two dimensional chromatography, there will only be two types of chromatography, thin layer and paper and all the other forms of chromatography will fall under the head of column chromatography. This is too simple a classification and does not take into consideration the nature of the mobile phase and the process responsible for separation. Moreover, two dimensional chromatography is operated using the liquid only as a mobile phase.

If the classification is done purely on the basis of the nature of mobile phase, there will be three types of chromatographic techniques – liquid, gas and supercritical fluid chromatography. Here, the first two are well known and several advancements have already been made in these two forms of chromatography. Supercritical fluid chromatography is relatively new and does not figure so frequently. Inspite of some advantages it has still to become popular. Again, in this type of classification, a large number of chromatographic techniques with varied modes of separations fall under the head of liquid chromatography.

In case the mechanism responsible for separation is taken as the criteria for classification, there will be four distinct types of chromatography viz, adsorption, partition, ion-exchange and size exclusion. But we will have to separately mention the competing phases.

No single criteria by itself will be able to effectively classify the various chromatographic techniques. We will have to take one as the main criteria and make a sub-classification based on the other yardstick. A simple approach seems to be to take the nature of the mobile phase as the main criteria and propose a sub-classification based on the various mechanisms responsible for separation. If any special feature exists within a sub-classification it may be pointed out right there.

For summing up the classification of chromatographic techniques, it emerges that there will be three types of chromatographic techniques:

- Liquid chromatography.
- Gas chromatography.
- Supercritical fluid chromatography.

Liquid Chromatography

The simplest one with liquid as a mobile phase seems to be based on partition between the liquid supported on an inert support and the other liquid acting as a mobile phase. This is known as liquid- liquid partition chromatography. The support holding the liquid can be in the form of three dimensions (column) or two dimensions (thin layer and paper). In the two dimensional form, the mobile phase can move from top to bottom or bottom to top and thus, known as descending or ascending chromatography. The two types of partition chromatography are also distinguishable on the basis of relative polarities of the mobile and stationary phases. Conventionally, the highly polar solvent is held on the support and the mobile phase is relatively non-polar. This is known as normal-phase chromatography. If the stationary phase is non-polar and the mobile phase relatively polar, it is termed as reversed-phase chromatography. An important modification in terms of stationary phase is introduced by loading the extractant used for solvent extraction on a hydrophobized inert support and irrigating the support with aqueous solvents. This is known as extraction chromatography. At this point, it may be important to introduce liquid chromatography on bonded phases. Initially, the partition chromatography was only liquid-liquid type but now chromatography with bonded phase has gained prominence.

In liquid-liquid type, a liquid stationary phase is held on the support packing by physical adsorption while in bonded phase the stationary phase is bonded chemically to the support. One of the main disadvantages of the liquid-liquid system is the loss of stationary phase by dissolution in the mobile phase. This requires periodic recoating of the support particles. Furthermore, the solubility of the stationary phase prohibits the use of liquid-phase packing for gradient elution. However, with bonded-phase, these problems are mitigated to a large extent. In a majority of cases, the bonded- phase packings for partition chromatography are prepared from rigid silica or silica based compositions. Bonded phase packings are classified as reversed phase when the bonded coating is non-polar and as the normal phase if the coating contains polar functional groups. Mostly in reversed phase the coating is a C_8 chain (n-octyl) or a C_{18} chain (n-octyldecyl). For normal phase polar bonded phases that have a diol ($-C_3H_6OCH_2CHOHCH_2OH$), cyano ($-C_2H_4CN$), amino ($-C_3H_6NH_2$) and dimethylamino ($-C_3H_6N(CH_3)_2$) functional groups are commercially available. The nature of the stationary phase and the way it is held on the support are responsible for quite a few diversifications in liquid partition chromatography. If the solid support instead of holding the liquid acts as an active adsorbent, the related chromatography is known as liquid-solid adsorption chromatography. This statement is true for both column and two dimensional chromatography. Sometimes, it becomes difficult to distinguish whether it is the partition or the adsorption which is dominating in the separation.

Advancement in liquid adsorption chromatography was reported in the form of affinity chromatography. In this the column material, a gel like material, is attached with a biospecific material through a spacer arm. This biospecific material is known as a ligand. The ligand selectively picks up a biological molecule from the mixture to form a complex. The desired biological molecule can be eluted by modest changes such as pH or ionic strength in the eluant. An important branch of liquid chromatography is ion exchange chromatography. This involves exchange of ions, cations or anions, between an insoluble matrix and the solution which comes in contact with it. The insoluble material which carries exchangeable cations and anions is known as ion exchanger. Those carrying exchangeable cations are known as cation exchangers and those with exchangeable anions are known as anion exchangers. There are amphoteric exchangers which are capable of both cation and anion exchange. Ion exchange, with a very few exceptions, is a reversible process. In contrast to adsorption, ion exchange is a stoichiometric process. Every ion removed from the solution is replaced by an equivalent amount of another ionic species of the same sign. Ion exchangers are organic and inorganic, natural and synthetic. Most of the ion exchangers in daily use are organic resins.

There is another form of chromatography which is synonymous to ion- exchange chromatography and is known as ion-pair (or paired-ion) chromatography. It is a type of reversed phase chromatography and used for the separation of ionic species. In this, the mobile phase consists of an aqueous buffer containing an organic solvent (methanol/ acetonitrile) and an organic compound containing a counter ion of opposite in charge to the analyte. The counter ion combines with the analyte to form a neutral ion pair. The neutral ion pair is retained on the non- polar stationary phase. It is subsequently eluted with an aqueous solution of methanol or other water soluble organic solvents. Ionic or partial ionic compounds can be chromatographed on reversed phase columns by using ion pairing reagents. Typically these reagents are long-chain alkyl anions or cations. When these reagents are used in dilute solutions, the retention of the analytical ion can increase. C_5 to C_{10} alkyl sulphonates are commonly used cationic compounds. Tetra alkyl ammonium salts (tetramethyl-, tetrabutyl-, and ammonium salt or triethyl- (C_5 - C_8) alkyl ammonium salts are generally used in the case of anionic solutes.

The applications of ion pair chromatography frequently overlap with those of ion exchange chromatography. Ion chromatography is generally used for the separations which are difficult to attain with ion exchange chromatography. A typical example is the separation of chlorate and nitrate which is difficult to achieve on an ion exchange but easily attained by ion pair chromatography.

Size exclusion chromatography is a very useful form of liquid chromatography employed for the separation of high molecular weight substances. This is also called gel permeation or gel filtration chromatography. Two important types of column packing material used for chromatography in aqueous phase are cross-linked dextrans and polyacrylamide. The process responsible for the separation is due to the distribution of solute between the aqueous phase with in the gel particles and external water. The selectivity is achieved on the basis of pore size of the gel. A solute which is small enough may enter the gel phase. In a column, this will have retarding effect. The molecules which have penetrated the gel, will spend their time sheltered from the moving phase. At the other extreme if the molecule is large enough not to be able to enter the pores will spend all the time in the mobile phase and move rapidly through the column. In between there will be molecules of intermediate size which can penetrate in some degree and their movement down the column will be somewhat retarded. This forms the basis for separating the molecules of different sizes.

The most significant advancement in liquid chromatography came in the form of high performance or high pressure liquid chromatography. This improved version is more or less available for different forms of liquid chromatography. The separation is performed on a column under pressure. The column usually contains particles of very small size, 3 -10 μm. The material may be an adsorbent or materials that permit partition, ion exchange or molecular permeation. The pressure generally used is around 6000 psi. The mobile phase may be composed of one solvent or a mixture of solvents with or without modifiers. Now high pressure thin layer chromatography (HPTLC) has also gained prominence.

Gas Chromatography

In gas chromatography, the mobile phase is a gas. The gas does not react with the molecules of the analyte. It only carries the sample through the column. Two types of gas chromatography are encountered:

- Gas-solid chromatography (GSC).
- Gas-liquid chromatography (GLC).

In GSC, the retention of the analyte is due to physical adsorption on the column material. In GLC, the separation takes place because of partition between the gaseous mobile phase and the liquid immobilized on the surface of an inert solid. GSC is not as popular as GLC because of some limitations of the former mode. In GSC, the retention time may be unusually large and high temperatures may be required to desorb the species. There may be severe tailing due to non-linear character of adsorption. The active adsorbent may catalyze the adsorbed species and change it to some other compound. As a result of all these GSC has a limited utility and gas chromatography more or less refers to gas-liquid partition chromatography. Generally, two types of columns are used in gas chromatography, packed and open tubular or capillary columns. The chromatographic columns vary in length from less than 2m to 50m or more. They are constructed of stainless steel, aluminium glass or fused glass or teflon. Open tubular columns are two basic types, wall-coated open tubular and support-coated open tubular. Wall coated columns are capillary tubes coated with thin layer of stationary phase. In support coated open tubular columns the inner surface is lined with a thin film of support material. This support material holds the stationary phase. The most commonly used stationary phases are polydimethyl siloaxanes. Organic species bonded to a solid surface are also used as stationary phases.

Supercritical Fluid Chromatography During the last quarter of the twentieth century, another type of chromatography with an entirely new type of mobile phase, supercritical fluid, has come up on the forefront. It is basically partition chromatography. Here, instead of a liquid or a gas, we use a supercritical fluid. In order to understand a supercritical fluid we will have to know critical temperature and pressure of a substance. The critical temperature of a substance is the temperature above which a distinct liquid phase cannot exist regardless of pressure. The vapour pressure of a substance at its critical temperature is its critical pressure. At temperature and pressure above its critical temperature and pressure (critical point), a substance is known as supercritical fluid. The properties like density, viscosity and other characteristics of supercritical fluids are intermediate between the gaseous and liquid state. The most well-known compound used as a supercritical fluid in chromatography is CO_2. The other compounds in use are nitrous oxide, ammonia and n- butane.

There are about two dozen compounds which have been used as mobile phases in supercritical chromatography. The critical temperature and pressure at these temperatures is well within the operating conditions used in HPLC.

One remarkable property of supercritical fluids which is related to their high densities (0.2- 0.5 g/ cm^3) is their capability to dissolve large non-volatile molecules. Another property which is associated with many supercritical fluids is that the dissolved analyte can be easily recovered by allowing the solutions to equilibrate with the atmosphere at a relatively low temperature. This is important if the analyte happens to be thermally unstable. A significant point which goes in the favour of many supercritical fluids is that they are inexpensive and non- toxic. They can be easily allowed to evaporate in atmosphere with no harmful effects.

As a matter of fact supercritical fluid chromatography is a hybrid of gas and liquid chromatography. It combines some of the advantages of both liquid and gas chromatography. In supercritical fluid chromatography, the mobile has a low viscosity and high self-diffusion coefficient. The mobile phase can be further modified by addition of polar modifier in small amounts. In this type of chromatography both packed and open tubular columns are used. In open tubular columns, the internal coating is of siloaxanes of various types and that of bonded phases.

Like gas chromatography, supercritical fluid chromatography is faster than liquid chromatography because of lower viscosity. Diffusion rates in supercritical fluid chromatography are intermediate between those in gases and in liquids. As a result of this, the band broadening is greater in supercritical fluids than in liquids but lesser than in gases. In nutshell, the intermediate diffusivities and viscosities of supercritical fluids may lead to faster separations than are obtained with liquid chromatography accompanied by lower zone broadening than observed in gas chromatography.

This new type of chromatography has started finding applications for a wide variety of materials like natural products, foods, pesticides, polymers, fossil fuels and explosives.

Elution of Columns

After having learnt the classification of chromatographic methods, it is important to look into some concepts which may explain the separation process. From the discussion, it can be noted that the separations are conducted on columns or on planar surfaces. The equilibria upon which the two types of chromatography are based are identical. The theory developed for column chromatography is adapted to planar chromatography.

To start with we look into a simple picture as to how the two substances X and Y are separated on a column by elution with a liquid as a mobile phase. Actually, elution involves washing a species through a column by continuous addition of fresh solvent. The different stages of separation are schematically shown in below fig. The sample is introduced at the head of column (time t_0) where the components of the sample distribute themselves between the two phases. Further addition of mobile phase pushes the solvent containing a part of the sample down the column. Here further partition between the mobile phase and fresh portion of stationary phase occurs (time t_1). Simultaneously, partitioning between the fresh solvent and the stationary phase is taking place at the site of the original sample.

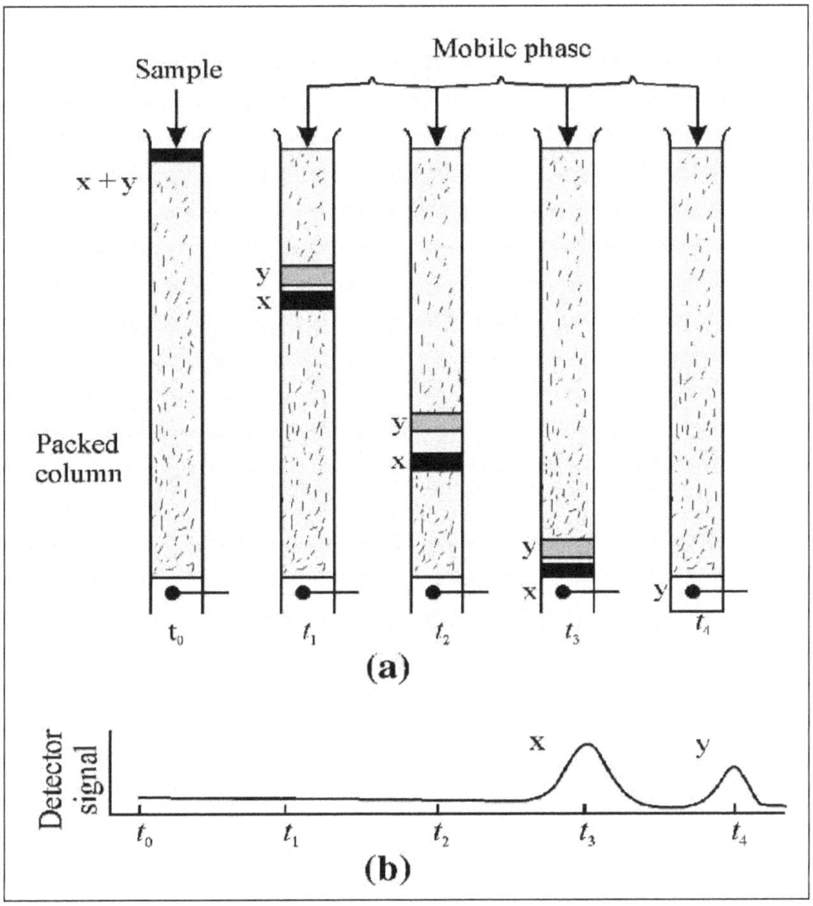

Fig: (a) Column separation of a mixture of components X and Y; and (b) Output of signal detector at various stages.

In this way, continued addition of solvent carries the solute molecules down the column in a continuous series of transfers between the mobile and stationary phases. The solute movement can only occur in the mobile phase; therefore, the average rate at which a solute zone migrates down the column depends upon fraction of time it spends in that phase. This fraction of time spent is small for solutes which are strongly retained by the stationary phase. If we refer to above Fig, this is the situation for Y. On the other hand, the retention of X is more in the mobile phase. As a result of this difference in retention rates, the components of mixture separate into bands or zones, (time t_2). If sufficient quantity of solvent is passed, the two species are isolated as individual zones and then pass out of the column where they can be detected or collected. (time t_3 and time t_4).

During the separation procedure, the dilution of analyte takes place. The analytes get more dilute than they were present in the original mixture. Therefore, the detectors employed for the separated components should be invariably more sensitive than would be required if the separation was not needed. The detector responds to the concentration of the solute coming out of the column. When its signal is plotted against time or the volume of the mobile phase a series of peaks are plotted. Such a plot is known as a chromatogram. The chromatogram is useful for both qualitative and quantitative analysis. The position of the peak on time axis helps to identify the components and the areas under the peak provide the amount of each component.

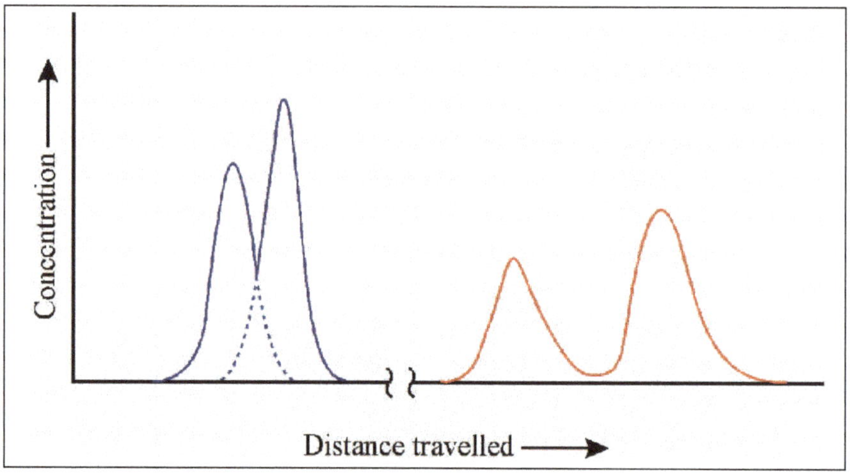

Fig: Concentration profiles at different stages of elution.

In the separation of X and Y, the species Y is more strongly held by the column and thus it lags. During the movement in the column the distance between the two zones increases. At the same time, broadening of two zones takes place. This lowers the efficiency of the separation method. There can be two approaches to increase the resolution of species:

- One component moves at a faster pace and the other at a slower rate.
- The rate of zone broadening for the two species is decreased.

The two situations are illustrated in figure below.

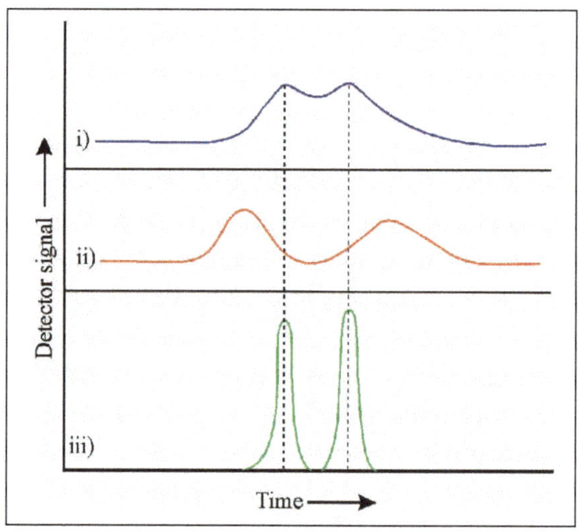

Fig: Two approaches to improve the binary separation; (i) initial chromatogram with overlapping peaks improved by; (ii) increase in band separation; and (iii) decrease in band width.

Migration Rates

From the above example, it becomes clear that the effectiveness of a column in separating the two solutes depends in part on the relative rate at which the two are eluted from the column. The rates depend upon the magnitude of the equilibrium constants of reactions by which the solutes distribute themselves between the mobile and stationary phases.

Distribution Constant

In chromatography, the distribution equilibrium is often simple and is described by the equations that involve the transfer of an analyte between the mobile and stationary phase. Thus, for the solute X, we may write:

$$X_{mobile} \Leftrightarrow X_{stationary}$$

The equilibrium constant, K, for the reaction is known as distribution constant and is given by the following expression:

$$K = \frac{C_S}{C_M}$$

Where, C_S is the molar concentration of the solute in the stationary phase and C_M is its molar concentration in the mobile phase. Ideally, C_S is directly proportional to C_M over a wide range of concentration. If such a situation exists, the chromatography is known as linear chromatography and results in symmetric Gaussian type peaks. The retention times are independent of the amount of the compound injected. Ideal behaviour, on the other hand, is not attainable in any actual chromatographic process. It is actually non-ideal behaviour which operates and the corresponding chromatography is known as non-linear chromatography.

Retention Time

It is time taken after sample injection for the analyte peak to reach the detector. It is denoted by the symbol t_R and in Fig below is represented.

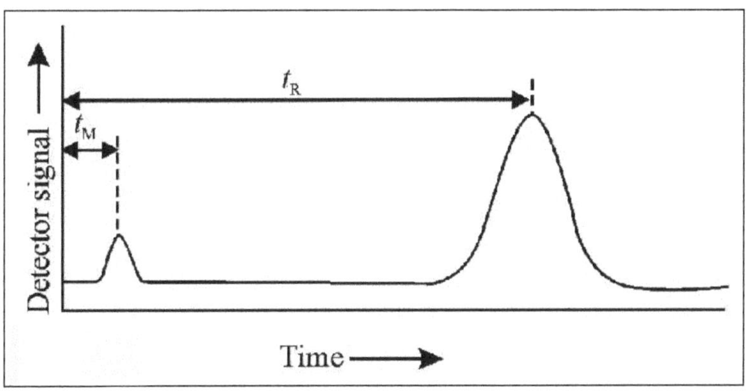

Fig: Small peak at the left represents a species that is not retained and reaches the detector immediately. t_M is approximately equal to the time needed for the molecules of the mobile phase to pass through the column.

The small peak in the beginning on the left is for the species that is not retained at the column. Sometimes, the sample may contain an un-retained species. If it is there it may be added because it helps in peak identification. The time t_M for the un-retained species is the same as the average rate of migration of mobile phase molecules. If L is the length of the column, the average rate of migration of solute molecules is given by the following expression:

$$v = \frac{L}{t_R}$$

Similarly, the average linear rate of migration of mobile phase molecule:

$$u = \frac{L}{t_M}$$

Where, t_M is called dead time. It is the time required for the average molecules of the mobile phase to pass through the column. After getting familiar with the terms distribution constant and retention time, it may be important to look into the relationship between the two terms. We express migration rate of a solute:

$$v = u \times \text{fraction of time solute spends in the mobile phase}$$

This fraction is equal to average number of moles of a solute in the mobile phase at any instant divided by the total number of moles of the solute in the column.

$$v = u \times \frac{\text{Moles of the solute in mobile phase}}{\text{Total moles in the column}}$$

The numbers of moles of the solute in the mobile and stationary phases are the product of their concentration and respective volumes:

$$v = u \times \frac{1}{1 + C_s V_s / C_M V_M}$$

Retention Factor

There is another term known as retention factor or capacity factor which is commonly used to describe the migration rates of the solutes. For a solute X, the retention factor ' X k is expressed as:

$$k_x = \frac{K_X V_S}{V_M}$$

K_X = distribution constant for X

$$v = u \times \frac{1}{1 + k_x}$$

Rearranging the above equation, we get:

$$k_x = \frac{t_R - t_M}{t_M}$$

We can determine t_R and t_M from a chromatogram. If the retention factor is much less than unity, elution occurs so rapidly that the accurate determination of retention factor is difficult. It the retention factor is larger, somewhere 20 or 30, the elution times are unusually long. Ideally separations of solutes are carried out for which the retention factors are between 2 and 10.

Selectivity Factor

The relative migration rates of the solutes are expressed in terms of selectivity factor. The selectivity factor/ separation factor, α, of a column for two solutes X and Y is expressed as follows:

$$\alpha = \frac{K_Y}{K_X}$$

Where, K_Y is the distribution constant for the more strongly retained species Y, and KX is the distribution constant for more rapidly eluted species X. The selectivity factor should always be greater than unity. Substituting $k_x = \frac{K_X V_S}{V_M}$ and the analogous equation for Y in Eq. $\alpha = \frac{K_Y}{K_X}$, the following equation is obtained for selectivity factor in terms of retention factor:

$$\alpha = \frac{k'_Y}{k'_X}$$

Where, k'_y and k'_x are retention factors for Y and X, respectively. Eq. $\alpha = \frac{k'_Y}{k'_X}$ can be converted into the form of retention times as follows:

$$\alpha = \frac{(t_R)_Y - t_M}{(t_R)_X - t_M}$$

Later, the selectivity factor and retention factor will be used to compute the resolving power of column.

Shapes of Peaks

At this stage, it may be important to examine the peaks in a chromatogram and the bands on a column. They show a similarity to normal error or Gaussian curves when the values of a measurement are plotted as a function of their occurrence. An individual solute molecule undergoes thousands of transfers between the stationary and mobile phase. In some cases, the residence time in a given phase may be transitory and in other case the period may be relatively long. The molecule is eluted during its residence in the mobile phase. Therefore, the migration is highly irregular down the column. Since there is variability in the residence time the average rate at which the individual molecules travel relative to mobile phase varies considerably. Some molecules travel faster because of their accidental inclusion in the mobile phase for a majority of time. On the contrary, others lag behind because they spend more time in the stationary phase. As a consequence of these random processes, a systematic spread of velocities around the mean value is observed. The breadth of band increases because more time is allowed for spreading to take place. It can be concluded that the zone breadth is directly related to the residence time in the column and inversely related to the velocity of the mobile phase.

Wastewater Treatment

The tremendous use of water in homes, factories, textile industry and a in numerous other works generates huge amount of wastewater. If it is discharge into water bodies without treatment, it will lead to sewer water pollution. It will pollute drinking water sources, rivers, oceans, ground water and even agricultural crops. It is therefore necessary that the efforts be made to treat the wastewater. There are several techniques, which are used in treatment of drinking and wastewater.

Sedimentation

Water obtained from different sources such as lake, river, sea, etc. always has suspended particles, which are produced by erosion of rock and soil. Whereas the large particle due to high mass settled down quite rapidly, the fine clay particle remains suspended for a long time. The suspended particles or sediments make the water turbid. In the treatment of wastewater, the suspended particles must be may settled down. Sedimentation is based on gravity to remove suspended solids from water. This process is also called settling or clarification.

Settling of solids depends on following factors:

- The gravitational force (with and without externally added chemicals) of the particle.
- Physical characteristics such as diameter, density, etc.
- Medium temperature, viscosity, density, etc.

Coagulation and Flocculation

Coagulation & Flocculation is necessary part of drinking water/ wastewater treatment. The water obtained from natural sources as well as wastewater contained small particulates which are colloidal in nature. They remain suspended and do not settle down under gravity. These colloidal particles carry the same charge. The repulsion between particles prevents them from combining into larger particulates to settle down. To precipitate these particles, the coagulation is carried out. Coagulation is commonly achieved by adding different types of chemicals (coagulants) to the wastewater to promote destabilization of the colloid dispersion and agglomeration of the resulting individual colloidal particles by neutralization of charge. Such compounds are called coagulants. In water treatment, the commonly used reagents as coagulants are ferrous sulfate ($FeSO_4.7H_2O$), aluminum sulfate ($Al_2\ (SO_4)_3.18H_2O$) and ferric chloride ($FeCl_3$).

Filtration

Filtration is the process of removing small particles from water by passing it through a porous medium. The small particles also include pathogens. The filtration techniques used are many. The most common type filter is used is granular media gravity filter.

In the pressure filter water is pumped under pressure. The filter media are artificial membranes, nets, sand filter, and high technological filter systems. The choice of filters depends on the required filtering speed and the cleanness requirement. The flow required for filtration can be achieved using gravity or pressure. In pressure filtration, one side of the filter medium is at higher pressure than that of the other so that the filter plane has a pressure drop. Some portion of this filter type must be enclosed in a container.

Water Disinfection

The drinking water must be free from harmful bacteria and viruses. It should also not contain fecal matter from animals and humans both. This is done by disinfection of water. The process requires an oxidant stronger than O_2. The different disinfection agents used are as follows.

Chlorination

It is done by passing chlorine in water. The process is cheaper and rapid and also most common. The chlorine reacts with water to produce hypochlorous acid at moderate pH.

$$Cl_2 + H_2O \rightarrow HOCl(aq.) + H^+ + Cl$$

The disinfectant is HOCl.

Ozonation

Ozone, which is a strong oxidant, is often used as a disinfectant. Ozone is produced at the site and bubbled through the water. It produces highly reactive radicals such as hydroxyl (OH) and hydroperoxy(HO_2). These destroy the harmful germs.

Wastewater Treatment is of three types: Primary, Secondary and Tertiary.

Primary Treatment

The purpose of primary treatment is to remove suspended and floating solids. This can be done by using both physical and chemical methods using some of the following steps:

- Screening: The different types of screens, viz., mesh and bar are used. The screens remove floating and suspended coarse particles when water is passed to screen.
- Sedimentation: In this process, the fine suspended particles which do not settle down under gravity due to their colloidal nature are coagulated by addition of alum etc. The coagulated particles are filtered out.
- Removal of Slit and Grit: The grit, which comprises sand and other heavier particles, is removed by slow passage of water through grit chambers.
- Removal of Greasy and Oily Matter: For this, water is kept in a skimming tank. When the compressed air is blown, oil and grease start floating on the water surface and are skimmed off.

Secondary Treatment

The secondary treatment of wastewater is mainly concerned with oxidation in order to purify wastewater.

- Biofiltration: This method utilizes different type of filters such as sand filters, contact filters, or trickling filters to remove the remaining sediment from wastewater.

- Aeration: In the wastewater after sedimentation process, activated sludge, which contains micro-organisms and aerobic bacteria, is mixed. Now the air is blown for aeration. The organic matter is oxidized under these aerobic conditions. This treated water is sent to a separate tank, where sludge is allowed to settle down. The purified water is collected.

Tertiary Treatment

This is the third and last step in the wastewater purification. It involves the following important steps:

- Removal of phosphates and ammonia: For removal of phosphate, lime is added whereby the following reaction takes place and calcium phosphate settles down as precipitate.

  $$3Ca(OH)_2 + 2PO_4^{3-} \rightarrow Ca(PO_4)_2 + 6OH^-$$

  $$NH4^+ + OH^- \rightarrow NH_3 + H_2O$$

- The last traces of suspended matter are removed by use of filter beds filled with gravels, coarse and fine sand.
- Disinfection: Harmful pathogenic bacteria are removed by using disinfectant such as chlorine, ozone and ultraviolet radiation.

Advanced Oxidation Processes for Wastewater Decontamination

Advanced Oxidation Processes (AOPs) and Advanced Oxidation Technologies (AOTs) are now used for wastewater decontamination. AOPs were first proposed by WH Glaze and coworkers for drinking water treatment with ozone, hydrogen peroxide and ultraviolet radiation.

Fundamentally, AOPs are based on the use of highly reactive radicals such as:

- Hydroxyl radicals ($OH^{\cdot}$).
- Sulfate radicals ($SO_4^{\cdot-}$).

These reactive species are the strongest oxidants that can oxidize almost any compound present in the water. These reactions are rapid and the rate constants are very high.

Properties of Hydroxyl Radicals

- It is highly reactive due to presence of an unpaired electron.
- It has very high redox potential ~ 2.89 volt (pH = 0) –1.95 volt (pH = 14).
- OH radical is very nonselective and rapidly reacts with numerous species at bimolecular rate constants, k, of the order of $\approx 10^8 - 10^{10}$ M^{-1} s^{-1}.
- These radicals have very short lifetime and are produced in situ.
- Hydroxyl radicals attack organic pollutants through four basic pathways:
  - Radical addition.

- Hydrogen abstraction.
- Electron transfer.
- Radical combination.

Basic Hydroxyl Reaction Mechanism

The most common reaction of hydroxyl radical is the abstraction of hydrogen atom. This initiates a radical chain oxidation:

$$RH + OH \rightarrow H_2O + R\ ,$$

$$2OH \rightarrow H_2O_2\,,$$

$$R + H_2O_2 \rightarrow ROH + OH\ ,$$

$$R + O_2 \rightarrow ROO\,,$$

$$ROO + RH \rightarrow ROOH + R\,.$$

Organic compounds produce carbon-centered radicals such as R or R–OH. These carboncentered radicals react with O_2 to form peroxyl radicals (ROO).

All these radicals further react and result in the formation of reactive species such as H_2O_2 and super oxide (O_2^-), leading to chemical degradation and even mineralization of organic compounds.

Table: Bimolecular rate constants ($M^{-1}\,s^{-1}$) of hydroxyl radicals.

Compounds	Rate Constants
Chlorinated alkenes	10^9–10^{11}
Phenols	10^9–10^{10}
N-containing organics	10^8–10^{10}
Aromatics	10^8–10^{10}
Ketones	10^9–10^{10}
Alcohols	10^8–10^9

Organic nitrogen is oxidized to nitrate /N_2 and organic sulfur is oxidized to the sulfate. Cyanide is oxidized to cyanate, which is then further oxidized to CO_2 and NO_3/N_2).

Production of Hydroxyl Radicals

Advanced oxidation methods for generation OH radicals employ several techniques.

Non-Photochemical

- Ozonation at elevated pH (>8.5).
- Ozone + hydrogen peroxide (O_3/H_2O_2).

- Ozone + catalyst (O_3/CAT).
- Fenton system (H_2O_2/Fe^{2+}).

Photochemical Methods

- O_3/UV.
- H_2O_2/UV,

  $$H_2O_2 + UV \rightarrow 2 \cdot OH.$$

- O_3/H_2O_2/UV.
- Photo-Fenton/Fenton-like system.
- Photocatalytic oxidation (UV/TiO_2)

  $$TiO_2 + UV \rightarrow e^- + h^+,$$

Irradiation of the photocatalytic surface leads to an excited electron (e−) and positive hole (h^+). The hydroxyl radical is formed by the reaction.

$$Ti(IV) + H_2O \rightleftharpoons Ti(IV)-H_2O,$$

$$Ti(IV)-H_2O + h+ \rightleftharpoons Ti(IV)-\cdot OH + H^+.$$

Precise, pre-programmed dosages, sequences and combinations of above reagents are used to obtain a maximum OH yield. In general, under appropriate conditions, AOPs are able to reduce the concentration of contaminants from several hundred ppm to less than 5 ppb. Obviously, AOPs bring out a significant reduction in COD and TOC of wastewater. AOPs are now called as water treatment processes of the 21st century.

Applications of Advanced Oxidation Processes

UV, UV/H_2O_2, or O_3/H_2O_2 systems are being used worldwide.

- For treating a variety of contaminants in process wastewater, groundwater, drinking water.
- Treatment of Well water by the O_3/H_2O_2 (ozone/peroxide) AOP for trichloroethylene.
- (TCE) and tetrachloroethylene (PCE) removal at a scale of 450 m^3/h in Los Angeles.
- An O_3/H_2O_2 system has been used for atrazine removal from the water of the River Seine in Paris at a scale of 5000 m^3/h, etc.
- UV disinfection is found in drinking water treatment plants throughout Europe and USA.

Many bacteria and viruses are inactivated and many organic compounds undergo photolysis in the presence of UV radiation.

Water Electrolysis

Electrolysis of water is a popular method used for different applications in various industries, mainly in the food industry, metallurgy, power plants amongst others. Besides, the components of water which include hydrogen and oxygen have many applications. For instance, hydrogen obtained through electrolysis is a clean, renewable and efficient fuel source.

The electrolysis of water is mainly carried out to yield pure hydrogen and oxygen gases. It involves passing an electric current through the water which results in the decomposition of water into hydrogen and oxygen.

However, the electrolysis of water is not simple and easy for many reasons:

- Water is very weakly dissociated into hydrogen and hydroxide ions. The concentration of the ions in neutral water are equal (= moles per litre). Electrolysis involves the charge carriers, for the current to flow. So, water with a very small amount of ions is a bad conductor of electricity. So, electrolysis of pure water will be a very slow process.
- The hydrogen ion is associated with other water molecule and exists as hydronium ion. So, any hydroxide ion, moving towards the anode, will be neutralized by the hydronium ion, even before it reaches the anode to form oxygen gas. Similarly, any hydrogen present will be neutralized by the hydroxyl ion present near the cathode and will not be, reduced to hydrogen. So, electrolysis of water to hydrogen and oxygen will be very small. The electrolysis also involves the transfer of electrons from the anion to the anode and cathode to cations.
- In the electrolysis of water, electrodes are inert solids like platinum/palladium whereas electrolyte is a solute in a solution and the product is a gas.

Factors Affecting the Efficiency of Electrolysis

The efficiency of electrolysis or the electron transfer depends on many factors such as:

- The number of available cations and anions in the solution.
- Mobility rate of the ions to reach the electrode.
- Activation energy needed for the electron transfer from the electrode to the electrolyte ions.
- The effect of the gas bubble surrounding the electrode on the further electrotransfer etc.

Crossing over of several interfaces (solute- liquid, solute-solid, solid-gas) results in the increase of energy requirements for the electrolysis (overvoltage) than predicted by the thermos-dynamical Gibbs energy.

Electrolysis of Water – Cell Potential and Thermodynamic Feasibility

Half reactions in the electrolysis of pure water at pH=7, and at 25°C are:

- At cathode: $2H_2O(l) + 2e^- \rightarrow H_2(g) + 2OH^-$ $E° = -0.42$ V

- At anode: $2H_2O \rightarrow O_2(g) + 4H^+ + 4e^-$ $E° = +0.82$ V
- The net reaction of electrolysis of water is given as $2H_2O(l) \rightarrow 2H_2(g) + O_2(g)$ $E° = -1.24$ V

The cell potential of electrolysis of pure water is negative and hence is thermodynamically unfavourable. Because of the low concentration of ions and the interfaces to be crossed electrons an extra voltage (Overvoltage) at each electrode is needed to about 0.6V.

In practice, continuous electrolysis of pure water is possible only at an external voltage of 2.4V. Since the electrolysis of pure water is thermodynamically non-feasible, methods to make it kinetically feasible are being investigated.

One of the methods is to increase the conductivity by increasing the number of ions available by adding acid, base, or non-reacting salts.

Water Electrolysis in the Presence of Acids with pH Lower than 7

Additional hydrogen ions from acid will be reduced at the cathode while water will be oxidized at the anode. Half reactions in an acid medium are:

- At cathode: $2H^+ + e^- \rightarrow H_2$ $E° = +0.0$ V
- At anode: $2H_2O \rightarrow O_2(g) + 4H^+ + 4e^-$ $E° = +1.23$ V
- Net reaction is written as $2H_2O \rightarrow O_2(g) + 2H_2$ $E° = -1.23$ V

The electrolysis takes place at a much lower potential than pure water (2.4V).

Water Electrolysis in the Presence of a Base with pH Higher than 7

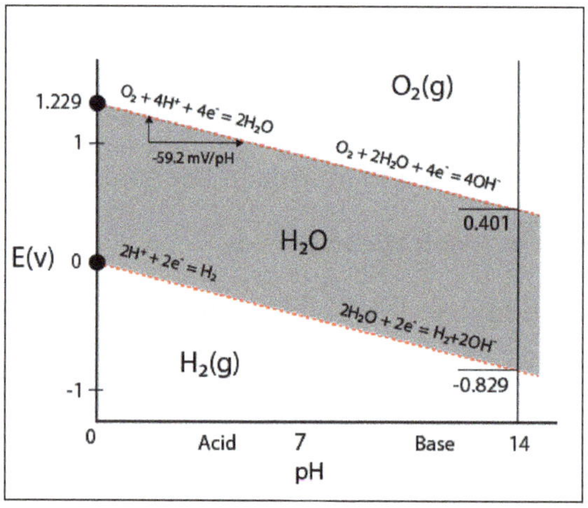

Additional hydroxyl ions, release their electrons to anode, while electrons at cathode oxidize water molecules near it. Half reactions of electrolysis in the presence of a base are:

- At cathode: $2H_2O(l) + 2e^- \rightarrow H_2(g) + 2OH^-$ $E° = -0.83$ V
- At anode: $4OH^- \rightarrow O_2 + 2H_2O + 4e^-$ $E° = +0.4$ V

- Net reaction is $2H_2O \rightarrow O_2(g) + 2H_2\ E° = -1.23\ V$

Like electrolysis in acid medium, electrolysis in the basic medium also needs much lower potential.

Pourbaix diagram gives the equilibrium regions of water, hydrogen and oxygen at various electrode potentials.

Water Electrolysis in the Presence of Salts

Salts are 100% dissociate into cations and anions in water and hence increase the ionic concentration for increasing conductivity. But the cations and anions from the salt also will be attracted towards the electrodes and hence become competitors to the decomposition of water to produce hydrogen and oxygen. So, the selection of salts with non-competing ions becomes necessary.

Salts containing lesser standard electrode potentials than hydrogen and hydroxide ions are suitable for the electrolysis of water.

Ions of first and second group elements (Li, Na, K, Mg, Ca, Ba, etc.) have lower standard potential than hydrogen ions and will not be reduced and allow hydrogen ions from water to hydrogen.

Non-reactive anions like nitrate, sulphate ions have lesser standard reduction potential than hydroxide ions. Sulphate oxidation to peroxy-sulphate has a reduction potential of +2.1V.

Non-soluble, solid polymeric ionic compounds (Nafion), has been found to help electrolysis of water in less than 1.5V.

Electrolysis of Water Using Electro Catalysts

Electro-catalysts are substances that accelerate electrochemical reactions without being consumed in the reaction like a catalyst in chemical reactions. Catalysts take the reaction through a different path of lower activation energy. High surface area, larger activation centres are the ability of the catalyst in increasing the reactivity.

The activity of the inert electrode like platinum can be, enhanced by modification of the surface by:

- Increasing the surface area with nanoparticles or alloying with catalytic d-block elements and changing the electronic state coated with other catalytic substance to enhance the electrolysis.
- Coating the electrode surface with catalytically active substances, like enzymes.

Electrolyzers

The electrolytic cell used for the electrolysis of water is the electrolyzer. Depending on the transporter of the electrolyte, electrolyzer can be divided into three types:

Polymer Electrolyte Membrane Electrolyzer

A polymer such as Nafion separates the electrodes and allows hydrogen ions formed by the oxidation of water at the anode to pass through it to the cathode compartment for discharge and form hydrogen gas.

Alkaline Electrolyzers

Dilute aqueous sodium (or potassium) hydroxide used in the electrolysis provides and movement of hydroxide ions to the anode to form oxygen.

Solid Oxide Electrolyzer

Ceramic oxide separates the electrodes. At the cathode, water is reduced to hydrogen and oxide ions. The oxide ions pass through the ceramic oxide to the anode to become oxygen gas. This is used at high temperatures of 700 to 800°C to reduce the external voltage needed for electrolysis.

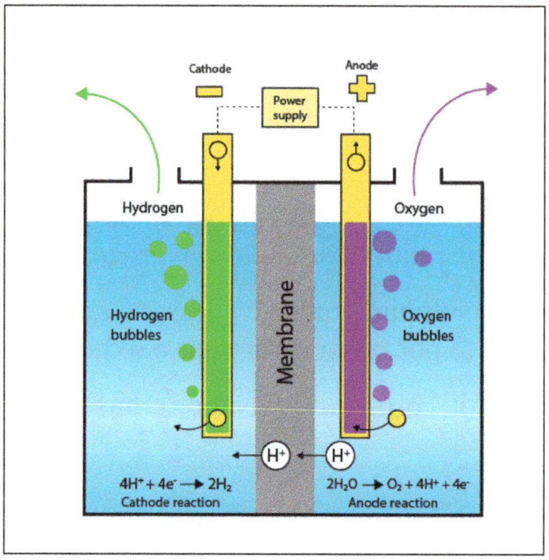

References

- Water-cycle: conserve-energy-future.com, Retrieved 29, May 2020
- Organic-pollutants, pollution: worldoceanreview.com, Retrieved 18, February 2020
- Water-pollution-by-heavy-metal-and-organic-pollutants-Brief-review-of-sources-effects-and-progress-on-remediation-with-aquatic-plants-337224742: researchgate.net, Retrieved 01, August 2020
- Water-electrolysis: byjus.com, Retrieved 12, March 2020

4

Anthroposphere and its Chemistry

The part of the environment which has been constructed or changed by the humans for their habitat and use comprise the anthroposphere. A few areas of study within this field are anthropogenic activities, industrial ecology and green chemistry. This chapter closely examines these key concepts of anthroposphere to provide an extensive understanding of the subject.

Anthroposphere

The anthroposphere encompasses the total human presence throughout the Earth system including our culture, technology, built environment, and associated activities. The anthroposphere complements the term anthropocene – the age within which the anthroposphere developed. Some mark this age as beginning with the advent of agriculture, others with the industrial revolution. A growing movement within the geological community is considering establishing the anthropocene as a new geologic era, possibly starting around 1950. In physical terms, the anthroposphere is comprised of the cities, villages, energy and transportation networks, farms, mines, and ports. It also encompasses books, software, blueprints, and communication systems - the mark of civilization.

Using a broad definition of the anthroposphere, it extends beyond Earth. For example, our radio and television broadcasts of the 20th and 21st centuries travel at the speed of light as an expanding electromagnetic sphere of human origin into the Milky Way galaxy. Another more tangible example are the four NASA probes, Voyagers 1 and 2 and Pioneers 10 and 11, launched in the 1970s and now venturing beyond our Solar System. In the event of an encounter with any other intelligent life, each probe contains images and other artifacts that attempt to convey the basic characteristics of humans and their niche within the universe.

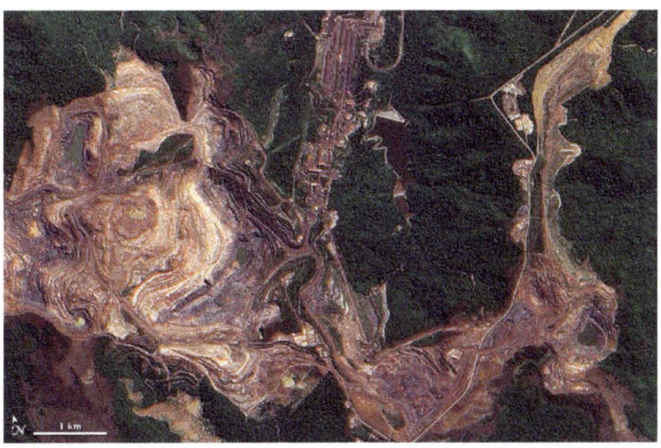

Carajàs Mine, Brazil: Nestled deep in remote Amazonia, the Carajàs mine is one of the world's largest iron ore mines. In a given year, this mine can produce 296 million metric tons of iron ore. Images like these reveal the vast amounts of resources from nature to supply their ever-growing needs, as well as the capability of humans to reshape the land.

Human Activities

Significant human (anthropogenic) activities include:

- Conversion of land to agriculture;
- Food production;
- Energy generation;
- Urban sprawl;
- Infrastructure development;
- Tropical deforestation;
- Transport extension;
- Aviation;
- Road transport;
- Harvesting natural resources for industrial production;
- Mining;
- Logging;
- Trawling.

Perspectives on the Anthroposphere

A key perspective of the Anthroposphere is of six component systems aggregated into three major sub-systems, representing human life on Earth:

The human system, comprising of:

- Individual Development: (civil liberties and human rights, equity, individual autonomy and self-determination, health, right to work, social integration and participation, gender and class-specific role, material standard of living, qualification, specialization, adult education, family and life planning horizon, leisure and recreation, arts);

- Social System: (population development, ethnic composition, income distribution and class structure, social groups and organisations, social security, medical care, old age provisions);
- Government: (government and administration, public finances and taxes, political participation and democracy, conflict resolution (national, international), human rights policy, population and immigration policy, legal system, crime control, international assistance policy, technology policy).

The support system, comprising of:

- Infrastructure: (settlements and cities, transportation and distribution, supply system (energy, water, food, goods, services), waste disposal, health services, communication and media, facilities for education and training, science, research and development);
- Economic system: (production and consumption, money, commerce and trade, labour and employment, income, market, inter-regional trade).

The natural system, comprising of:

- Resources and Environment: (natural environment, atmosphere and hydrosphere, natural resources, ecosystems, species, depletion of non-renewable resources, regeneration of renewable resources, waste absorption, material recycling, pollution, degradation, carrying capacity);

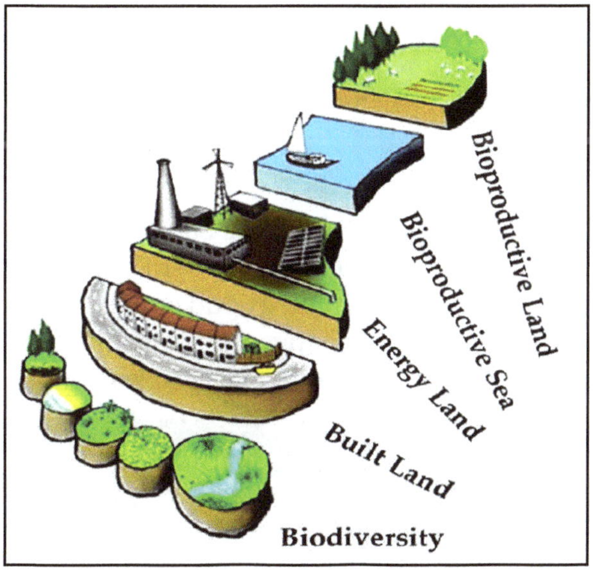

Separate Sphere for Humans

For much of our existence, human impact on the environment was comparable to other mammals roaming the Earth. As our numbers grew and our impact on the landscape expanded, we departed from the trajectory common to other mammals. Harnessing fire and tool-making, followed by the advent of agriculture and sustained settlements, are two critical departures from those commonalities.

Since the start of the industrial revolution in the late 18th century, humans developed the ability to harness the power of fossil fuels, transcending our role from observers to drivers of global change. Human dwellings now occupy about 8 percent of ice-free land, but about three-fourths of the land surface has been altered by humans in some way. Additionally, we have significantly altered the biogeochemical cycles of nitrogen, carbon, sulfur, and phosphorus. Alteration of the carbon cycle alone has changed the pH of the oceans and the climate of Earth. Chemical inventions such as chlorinated fluorocarbons (CFCs) have altered the ozone in the stratosphere and the amount of ultraviolet light reaching Earth's surface. The footprint of our chemical activities is found in the air, water, land, and biota of Earth in the form of naturally occurring and human created molecules.

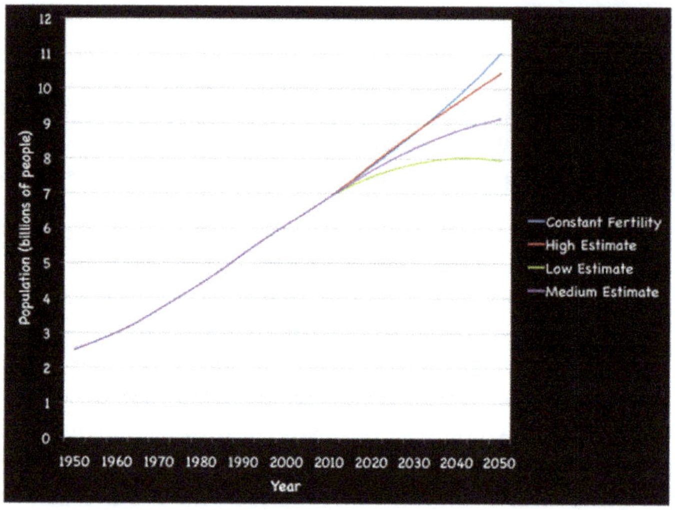

World Population: A key attribute of the anthroposphere over the past century has been rapid and sustained population growth. Estimates for world population show between 8 billion and 10.5 billion people living on the Earth by 2050.

Changes in Anthroposphere

When compared with most natural changes in other spheres, change in the anthroposphere is happening rapidly. This is partly due to the rapid increase in population over the past several centuries but also as a result of the strides in technology and energy that have empowered humans to directly and indirectly affect environmental change. In terms of world population, the number of humans has soared from about 1 billion in 1800 to over 7 billion today.

Humans are the only species whose presence is recognizable from space. While humans are distributed throughout most of the land, the contrast between the luminous and dark regions of the continents reveal disparities in human development that exist throughout the globe.

To make room for and to feed these added billions, the anthroposphere has expanded to occupy more land for dwellings and agriculture. In addition, human appropriation of fossil fuel energy and the many technologies it powers have played a major role in amplifying the influence of the anthroposphere in the Earth system. It has enabled humans to extract resources such as iron ore and bauxite, which are used to make up automobiles, skyscrapers, and countless gadgets integral to modern life. However, these activities have expanded the anthroposphere in subtle ways by infusing pollutants into our water and air, negatively affecting the biosphere and bringing about global warming.

A defining decision for humans this century will be to either maintain a business-as-usual course, which many experts estimate could cause widespread devastation to life on Earth, or alter the nature of human activities in order to insure greater balance between the anthroposphere and the rest of the Earth's systems.

Anthropogenic Activities

Southern approach to the Sydney Harbour Bridge, NSW.

Human activities have increased greenhouse gas concentrations in the atmosphere Atmospheric concentrations of carbon dioxide (CO_2), methane and nitrous oxide began to rise around two hundred years ago, after changing little since the end of the last ice age thousands of years earlier. The concentration of CO_2 has increased from 280 parts per million (ppm) before 1800, to 396 ppm in 2013. This history of greenhouse gas concentrations has been established by a combination of modern measurements and analysis of ancient air bubbles in polar ice.

Particularly important is CO_2. Enormous amounts of it are continually exchanged between the atmosphere, land and oceans, as land and marine plants grow, die and decay, and as carbon-rich waters circulate in the ocean. For several thousand years until around 200 years ago, this 'carbon cycle' was approximately in balance and steady. Since the 19th century, human-induced CO_2 emissions from fossil fuel combustion, cement manufacture and deforestation have disturbed the balance, adding CO_2 to the atmosphere faster than it can be taken up by the land biosphere and the oceans. On average over the last 50 years, about 25% of total CO_2 emissions were absorbed by the ocean making sea water more acidic and 30% was taken up on land, largely by increased plant growth stimulated by rising atmospheric CO_2, increased nutrient availability, and responses to warming and rainfall changes. The other 45% of emissions accumulated in the atmosphere. These changes to the carbon cycle are known from measurements in the atmosphere, on land and in the ocean, and from modeling studies.

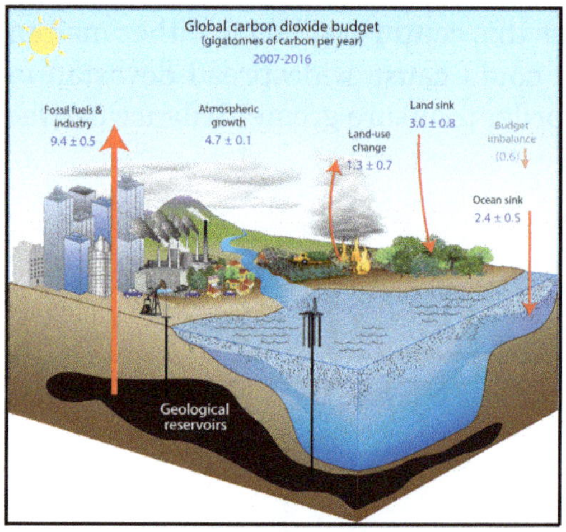

Figure: The natural carbon cycle, in which CO_2 circulates between the atmosphere, land and oceans, has been changed by emissions of CO_2 from human activities. In this diagram of the global carbon cycle, numbers on arrows represent carbon flows averaged over 2004–2013, in gigatonnes of carbon per year.

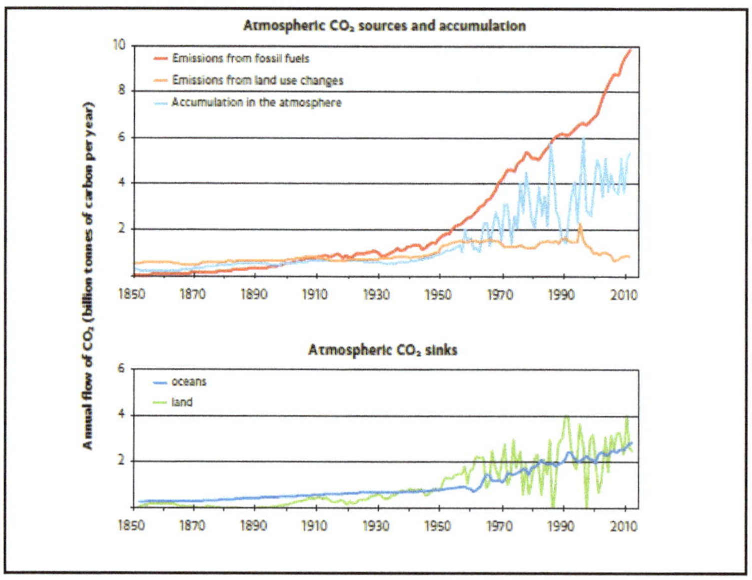

Figure: An 'atmospheric CO_2 budget' reveals the amount of carbon in the net amounts of CO_2 entering, leaving and accumulating in the atmosphere.

The dominant cause of the increasing concentration of CO_2 in the atmosphere is the burning of fossil fuels. Over the last two centuries, the growth of fossil fuel combustion has been closely coupled to global growth in energy use and economic activity. Fossil fuel emissions grew by 3.2% per year from 2000 to 2010, a rapid growth that is dominated by growth in Asian emissions and has exceeded all but the highest recent long-range scenarios for future emissions.

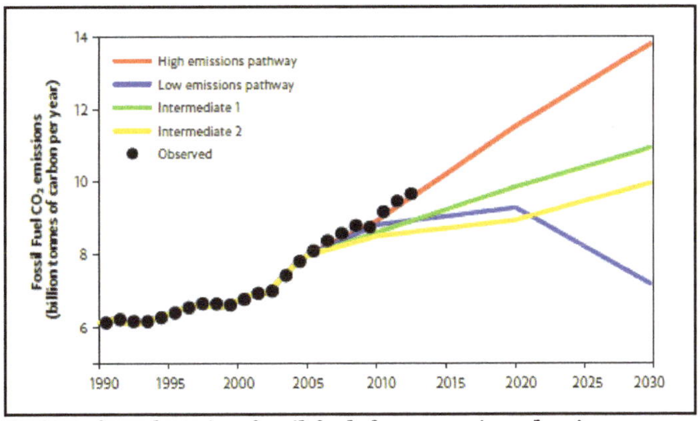

Figure: CO_2 emissions from burning fossil fuels have continued to increase over recent years.

Although fossil-fuel emissions of CO_2 have grown fairly steadily, the upward march of the CO_2 concentration in the atmosphere varies from year to year. This is caused mainly by the effects of weather variability on vegetation, and also by sporadic volcanic activity: major volcanic eruptions have a significant indirect influence on atmospheric CO_2 concentrations, causing temporary drawdown of CO_2 through the promotion of plant growth by the light-scattering and cooling effects of volcanic haze. By contrast, the direct contribution of volcanic emissions to atmospheric CO_2 is negligible, amounting to around 1% of current human induced emissions.

Wollongong, NSW at night.

Most of the Observed Recent Global Warming Results From Human Activities

Climatic warming or cooling arises from changes in the flows of energy through the climate system that can originate from a number of possible driving factors. The main drivers that have acted over the last century are:

- Increases in atmospheric CO_2 and other long-lived greenhouse gases.
- Increases in short-lived greenhouse gases.
- Changes to land cover.
- Increases in aerosols.
- Solar fluctuations.
- Volcanic eruptions.

Of these, solar fluctuations and volcanic eruptions are entirely natural, while the other four are predominantly caused by human influences. The human-induced drivers have been dominant over the past century. Changes in greenhouse gas concentrations, dominated by CO_2, caused a large warming contribution. Some of this has been offset by the net cooling effects of increased aerosol concentrations and their impact on clouds. Black carbon or soot has probably exerted a smaller, warming influence. The net effect of all aerosol types including soot remains hard to quantify accurately. Among the natural influences, the effect of changes in the brightness of the Sun has been very small. Volcanic influences are highly intermittent, with major eruptions causing significant cooling for a year or two, but their average effects over the past century have been relatively small.

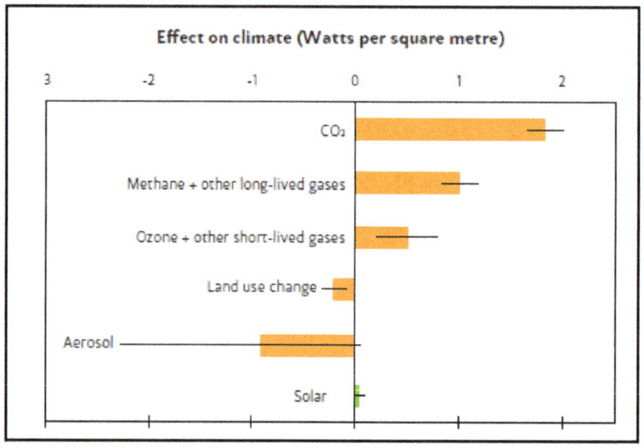

Figure: Human-induced drivers of climate change have been much larger than natural drivers over the last century. The strength of these drivers, which are changing the long-term energy balance of the planet, is measured in Watts per square metre.

Using climate models, it is possible to separate the effects of the natural and human-induced influences on climate. Models can successfully reproduce the observed warming over the last 150 years when both natural and human influences are included, but not when natural influences act alone. This is both an important test of the climate models against observations and also a demonstration that recent observed global warming results largely from human rather than natural influences on climate.

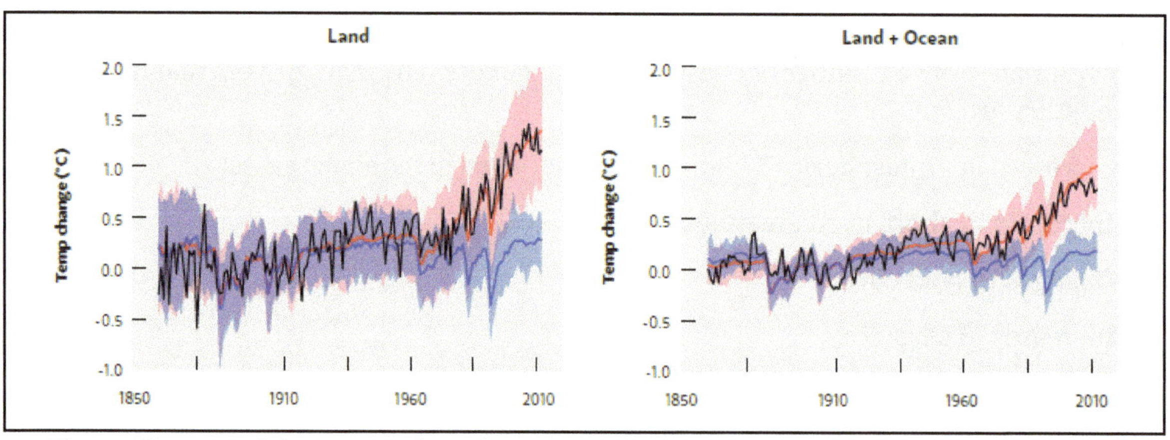

Figure: Climate models can correctly replicate recent warming only if they include human influences.

It is also possible to distinguish the effects of different human and natural influences on climate by studying particular characteristics of their effects. For example, it was predicted more than a century ago that increases in CO_2 would trap more heat near the surface and also make the stratosphere colder. In recent years, satellite and other measurements have provided strong evidence that the upper atmosphere has cooled and the lower atmosphere has warmed significantly—the predicted consequence of extra greenhouse gases. This supports the inference that the observed near surface warming is due primarily to an enhanced greenhouse effect rather than, say, an increase in the brightness of the Sun.

Rainforest canopy, Bellenden Ker Range, North Queensland.

Some Recent Changes in Australia's Climate are Linked to Rising Greenhouse Gases

Modeling studies indicate that rising greenhouse gases have made a clear contribution to the recent observed warming across Australia. Depletion of the ozone layer in the upper atmosphere over Antarctica and rising greenhouse gas concentrations are also likely to have contributed significantly to climate trends that have been observed in the Australian region over the past two decades. These include stronger westerly winds over the Southern Ocean, strengthening of the high-pressure ridge over southern Australia, and a related southward shift of weather systems. These trends are consistent with climate model projections, and are likely to be largely human-induced through a combination of increases in greenhouse gases and thinning of the ozone layer.

Past decadal trends in Australian rainfall cannot yet be clearly separated from natural climate variations, except in southwest Western Australia where a significant observed decline in rainfall has been attributed to human influences on the climate system.

There has very likely been net uptake of CO_2 by Australian vegetation, consistent with global uptake of CO_2 by vegetation on land. This has been accompanied by increases in the greenness of Australian vegetation, which is also consistent with global trends.

Industrial Ecology

Industrial ecology conceptualises industry as a man-made ecosystem that operates in a similar way to natural ecosystems, where the waste or by product of one process is used as an input into another process. Industrial ecology interacts with natural ecosystems and attempts to move from a linear to cyclical or closed loop system. Like natural ecosystems, industrial ecology is in a continual state of flux.

Main Features: Industrial processes, from material extraction through to product disposal, have an adverse impact upon the environment. Industrial ecology aims to reduce environmental stress caused by industry whilst encouraging innovation, resource efficiency and sustained growth. Industrial ecology acknowledges that industry will continue operate and expand however, it supports industry that is environmentally conscious and has less burden upon the planet. It views industrial sites as part of a wider ecology rather than an external, solitary entity.

Within the industrial ecology concept, industry interacts with nature and utilises the wastes and by products of other industries as inputs into its own processes. Industrial ecology ranges from purely industrial ecosystems to purely natural ecosystems with a range of hybrid industrial/natural ecosystems in between. Covering both industrial management and technology, industrial ecology encompasses other sustainability concepts and tools such as material flows analysis; environmentally sound technologies; design for disassembly; and dematerialisation.

The principles of industrial ecology as defined by Tibbs (1992) are:

- Create industrial ecosystems: close the loop; view waste as a resource; create partnerships with other industries to trade by-products which are used as inputs to other processes.
- Balance industrial inputs and outputs to natural levels: manage the environmental-industrial interface; increase knowledge of ecosystem behaviour, recovery time and capacity; increase knowledge of how and when industry can interact with natural ecosystems and the limitations.
- Dematerialisation of industrial output: use less virgin materials and energy by becoming more resource efficient; reuse materials or substituting more environmentally friendly materials; do more with less.
- Improve the efficiency of industrial processes: redesign products, processes, equipment; reuse materials to conserve resources.
- Energy use: incorporate energy supply within the industrial ecology; use alternative sources of energy that have less or no impact upon the environment.
- Align policies with the industrial ecology concept: incorporate environment and economics into organisational, national and international policies; internalize the externalities; use economic instruments to encourage a move towards industrial ecology; use a more appropriate discount rate; use a more comprehensive index to measure a nation's wealth rather than GNP.

The benefits of industrial ecology include: cost savings (materials purchasing, licensing fees, waste disposal fees, etc); improved environmental protection; income generation through selling waste or by products; enhanced corporate image; improved relations with other industries and organisations and market advantages. Limitations to industrial ecology include: no market for materials; lack of support from government and industry; reluctance of industry to invest in appropriate technology; perceived legal implications and reluctance to move to another supplier.

The formation of virtual or physical eco-parks arises from clusters of industry that agree to supply or sell waste to each other, thereby moving towards the industrial ecology concept. Most eco-parks are virtual due to the high cost associated with relocating facilities. However some physical eco-parks are being designed whereby certain industries are located on the same site.

Case Studies and Example

Industry Partnerships: Since the 1970's several industries in Denmark have supplied or sold by products and wastes to other industries. Asnaes, the largest coal-fired power plant in Denmark, sold processed steam to Statoil (an oil refinery) and Novo Nordisk (a pharmaceutical plant). Some of Asnaes' surplus heat was supplied to the town's heating scheme, reducing the number of domestic oil burning systems in use. Surplus heat was also used to heat the water of Asnaes' commercial fish farm. Local farmers used sludge from the fish farm as fertilizer. By treating some of its waste, Novo Nordisk sold high nutrient liquid sludge to farmers. Statoil supplied cooling and purified waste water to Asnaes which reduced Asnaes' freshwater extraction. In addition, Statoil removed sulphur from its surplus gas and sold all of its cleaned surplus gas to Asnaes and Gyproc (a plaster-board factory). The removed sulfur was sold to Kemira (a sulfuric acid producer). By desulfurising its smoke, Asnaes sold the resulting calcium sulfate to Gyproc as an alternative to mined gypsum which was being imported.

These partnerships were formed voluntarily and negotiated independently. Initially for purely economic reasons, some of the later deals were made for environmental reasons.

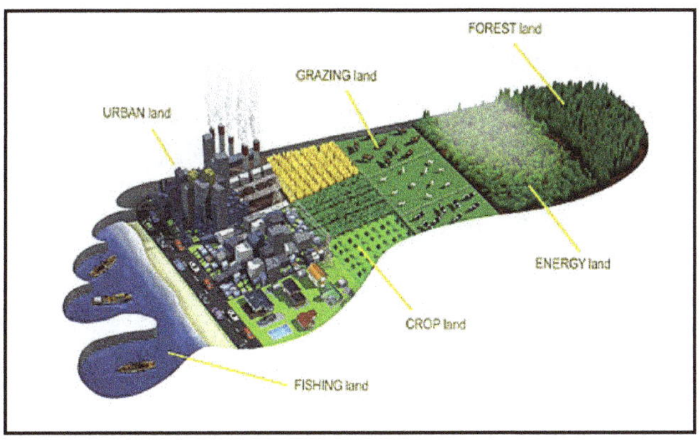

Waste Management

Americans alone are responsible for producing a hopping 220 million tons of waste a year. This number is far more than any other nation in the world. Because of this fact both the government and environmental associations have developed numerous methods of dealing with the problem. Waste management is that solution, a rather complex issue that encompasses more than 20 different industries. Waste management is collection, transportation, and disposal of garbage, sewage and other waste products.

Waste management is the process of treating solid wastes and offers variety of solutions for recycling items that don't belong to trash. It is about how garbage can be used as a valuable resource. Waste management is something that each and every household and business owner in the world needs. Waste management disposes of the products and substances that you have use in a safe and efficient manner.

"Waste management or Waste disposal is all the activities and actions required to manage waste from its inception to its final disposal. This includes amongst other things, collection, transport, treatment and disposal of waste together with monitoring and regulation. It also encompasses the legal and regulatory framework that relates to waste management encompassing guidance on recycling etc."

You will find there are eight major groups of waste management methods, each of them divided into numerous categories. Those groups include source reduction and reuse, animal feeding, recycling, composting, fermentation, landfills, incineration and land application. You can start using many techniques right at home, like reduction and reuse, which works to reduce the amount of disposable material used.

Various Methods of Waste Disposal.

Although there are many methods available to dispose off waste. Let's take a look at some of the most commonly used methods that you should know about waste management.

Landfills

Throwing daily waste/garbage in the landfills is the most popularly used method of waste disposal used today. This process of waste disposal focuses attention on burying the waste in the land. Landfills are commonly found in developing countries. There is a process used that eliminates the odors and dangers of waste before it is placed into the ground. While it is true this is the most popular form of waste disposal, it is certainly far from the only procedure and one that may also bring with it an assortment of space.

This method is becoming less these days although, thanks to the lack of space available and the strong presence of methane and other landfill gases, both of which can cause numerous contamination problems. Landfills give rise to air and water pollution which severely affects the environment and can prove fatal to the lives of humans and animals. Many areas are reconsidering the use of landfills.

Incineration/Combustion

Incineration or combustion is a type disposal method in which municipal solid wastes are burned at high temperatures so as as to convert them into residue and gaseous products. The biggest advantage of this type of method is that it can reduce the volume of solid waste to 20 to 30 percent of the original volume, decreases the space they take up and reduce the stress on landfills.

This process is also known as thermal treatment where solid waste materials are converted by Incinerators into heat, gas, steam and ash. Incineration is something that is very in countries where landfill space is no longer available, which includes Japan.

Recovery and Recycling

Resource recovery is the process of taking useful discarded items for a specific next use. These discarded items are then processed to extract or recover materials and resources or convert them to energy in the form of useable heat, electricity or fuel.

Recycling is the process of converting waste products into new products to prevent energy usage and consumption of fresh raw materials. Recycling is the third component of Reduce, Reuse and Recycle waste hierarchy. The idea behind recycling is to reduce energy usage, reduce volume of landfills, reduce air and water pollution, reduce greenhouse gas emissions and preserve natural resources for future use.

Plasma Gasification

Plasma gasification is another form of waste management. Plasma is a primarily an electrically charged or a highly ionized gas. Lighting is one type of plasma which produces temperatures that exceed 12,600 °F . With this method of waste disposal, a vessel uses characteristic plasma torches operating at +10,000 °F which is creating a gasification zone till 3,000 °F for the conversion of solid or liquid wastes into a syngas.

During the treatment solid waste by plasma gasification, the waste's molecular bonds are broken down as result of the intense heat in the vessels and the elemental components. Thanks to this process, destruction of waste and dangerous materials is found. This form of waste disposal provides renewable energy and an assortment of other fantastic benefits.

Composting

Composting is a easy and natural bio-degradation process that takes organic wastes i.e. remains of plants and garden and kitchen waste and turns into nutrient rich food for your plants. Composting, normally used for organic farming, occurs by allowing organic materials to sit in one place for months until microbes decompose it. Composting is one of the best method of waste disposal as it can turn unsafe organic products into safe compost. On the other side, it is slow process and takes lot of space and turns it to.

Waste to Energy

Waste to energy(WtE) process involves converting of non-recyclable waste items into useable heat, electricity, or fuel through a variety of processes. This type of source of energy is a renewable energy source as non-recyclable waste can be used over and over again to create energy. It can also help to reduce carbon emissions by offsetting the need for energy from fossil sources. Waste-to-Energy, also widely recognized by its acronym WtE is the generation of energy in the form of heat or electricity from waste.

Avoidance/Waste Minimization

The most easier method of waste management is to reduce creation of waste materials thereby reducing the amount of waste going to landfills. Waste reduction can be done through recycling old materials like jar, bags, repairing broken items instead of buying new one, avoiding use of disposable products like plastic bags, reusing second hand items, and buying items that uses less designing.

Recycling and composting are a couple of the best methods of waste management. Composting is so far only possible on a small scale, either by private individuals or in areas where waste can be mixed with farming soil or used for landscaping purposes. Recycling is widely used around the world, with plastic, paper and metal leading the list of the most recyclable items. Most material recycled is reused for its original purpose.

The Bottom Line

There are certain waste types that are considered as hazardous and cannot be disposed of without special handling which will prevent contamination from occurring. Biomedical waste is one example of such. This is found in health care facilities and similar institutions. The special waste disposal system for this unit in place to dispose of this type of waste.

As you can see there are plenty of important things that you should know about waste management and disposal in order to ensure that you are safe, as well as that you are keeping the environment safe. It is your choices as to how you will dispose of waste, however it is always in your best interest to take a look at all of the options that you have available before making the choice.

Industrial Waste

We have millions of factories, mills, industries, mining plants, etc. around the world. These industries use raw materials to produce finished goods for consumers. But in the manufacturing process, there are materials which are rendered useless. They constitute the industrial waste. Some examples of industrial wastes are metals, paints, sandpaper, slag, ash, radioactive wastes, etc.

Types of Industrial Waste:

Industrial waste can be categorized into biodegradable and non-biodegradable:

- Biodegradable: Those industrial wastes which can be decomposed into the non-poisonous matter by the action of certain microorganisms are the biodegradable wastes. They are even comparable to house wastes. These kinds of waste are generated from food processing industries, dairy, textile mills, slaughterhouses, etc. Some examples are paper, leather, wool, animal bones, wheat, etc. They are not toxic in nature and they do not require special treatment either. Their treatment processes include combustion, composting, gasification, bio-methanation, etc.
- Non-biodegradable: Those industrial wastes which cannot be decomposed into non-poisonous substances are the non-biodegradable wastes. Examples are plastics, fly ash, synthetic fibers, gypsum, silver foil, glass objects, radioactive wastes, etc. They are generated by iron and steel plants, fertilizer industries, chemical, drugs, and dyes industries. It is estimated that about 10 to 15 percentage of the total industrial wastes are non-biodegradable and hazardous, and the rate of increase in this category of waste is only increasing every year. These wastes cannot be broken down easily and made less harmful. Hence, they pollute the environment and cause threat to living organisms. They accumulate in the environment and enter the bodies of animals and plants causing diseases. However, with the

advancement in technology, several disposals, and reuse methods have been developed. Wastes from one industry are being treated and utilized in another industry. For example, the cement industry uses the slag and fly ash generated as waste by steel industries. Landfill and incineration are other methods which are being resorted to, for the treated of hazardous wastes.

Causes of Industrial Pollution

1. Lack of Policies to Control Pollution: Lack of effective policies and poor enforcement drive allowed many industries to bypass laws made by pollution control board which resulted in mass scale pollution that affected lives of many people.

2. Unplanned Industrial Growth: In most industrial townships, unplanned growth took place wherein those companies flouted rules and norms and polluted the environment with both air and water pollution.

3. Use of Outdated Technologies: Most industries still rely on old technologies to produce products that generate large amount of waste. To avoid high cost and expenditure, many companies still make use of traditional technologies to produce high end products.

4. Presence of Large Number of Small Scale Industries: Many small scale industries and factories that don't have enough capital and rely on government grants to run their day-to-day businesses often escape environment regulations and release large amount of toxic gases in the atmosphere.

5. Inefficient Waste Disposal: Water pollution and soil pollution are often caused directly due to inefficiency in disposal of waste. Long term exposure to polluted air and water causes chronic health problems, making the issue of industrial pollution into a severe one. It also lowers the air quality in surrounding areas which causes many respiratory disorders.

6. Leaching of Resources from Our Natural World: Industries do require large amount of raw material to make them into finished products. This requires extraction of minerals from beneath the earth. The extracted minerals can cause soil pollution when spilled on the earth. Leaks from vessels can cause oil spills that may prove harmful for marine life.

Effects of Industrial Pollution

1. Water Pollution: The effects of industrial pollution are far reaching and liable to affect the eco-system for many years to come. Most industries require large amounts of water for their work. When involved in a series of processes, the water comes into contact with heavy metals, harmful chemicals, radioactive waste and even organic sludge.

These are either dumped into open oceans or rivers. As a result, many of our water sources have high amount of industrial waste in them which seriously impacts the health of our eco-system. The same water is then used by farmers for irrigation purpose which affects the quality of food that is produced.

Water pollution has already rendered many ground water resources useless for humans and wild-life. It can at best be recycled for further usage in industries.

2. Soil Pollution: Soil pollution is creating problems in agriculture and destroying local vegetation. It also causes chronic health issues to the people that come in contact with such soil on a daily basis.

3. Air Pollution: Air pollution has led to a steep increase in various illnesses and it continues to affect us on a daily basis. With so many small, mid and large scale industries coming up, air pollution has taken toll on the health of the people and the environment.

4. Wildlife Extinction: By and large, the issue of industrial pollution shows us that it causes natural rhythms and patterns to fail, meaning that the wildlife is getting affected in a severe manner. Habitats are being lost, species are becoming extinct and it is harder for the environment to recover from each natural disaster. Major industrial accidents like oil spills, fires, leak of radioactive material and damage to property are harder to clean-up as they have a higher impact in a shorter span of time.

5. Global Warming: With the rise in industrial pollution, global warming has been increasing at a steady pace. Smoke and greenhouse gases are being released by industries into the air which causes increase in global warming. Melting of glaciers, extinction of polar beers, floods, tsunamis, hurricanes are few of the effects of global warming.

The issue of industrial pollution concerns every nation on the planet. As a result, many steps have been taken to seek permanent solutions to the problem. Better technology is being developed for disposal of waste and recycling as much polluted water in the industries as possible. Organic methods are being used to clean the water and soil, such as using microbes that naturally uses heavy metals and waste as feed. Policies are being pushed into place to prevent further misuse of land. However, industrial pollution is still rampant and will take many years to be brought under control.

Industrial Waste management

Industrial waste management is the set of strategies used to handle industrial waste removal and disposal in a timely but cost-effective manner that is also in compliance with regulations.

The Importance of Industrial Waste Management:

- Good for the Environment:

  Poor industrial waste management can lead to those wastes getting into the local environment and then causing possibly irreparable damage. For example, chemical pollutants from manufacturing and similar industrial processes can enter the water and get into the bodies of everything living in that water. Certain chemicals can build up if ingested, meaning that higher and higher concentrations build up in the bodies of animals higher on the food chain, including potentially humans. Proper industrial waste management will help deter such outcomes while also ensuring that less hazardous industrial wastes are recycled, which is something that comes with its own benefits.

- Protecting Human Health:

  Improper handling and disposal of industrial waste can contribute to increased incidents of serious medical conditions among the human residents of the local region. For example, if the run-off from a cattle ranch enters the local river, residents living down-river from that ranch can find themselves falling ill with diseases such as e-coli infections. Effective industrial waste management will not only do a better job of ensuring that such outcomes never happen, but also reducing the amount of industrial waste that is produced in the first place.

- Legal Liability:

  Businesses have the obligation to effectively manage and dispose of their industrial waste. If they fail to do so, they will become liable for the damages caused through their negligence. Even in cases where the businesses win their cases, they will still lose out on legal fees and time spent defending themselves in the courts.

Chemical Waste

Chemical waste is any type of waste that is composed of noxious, potentially hazardous chemicals. Harmful chemicals and solvents that are the byproducts of large scale laboratories and manufacturing plants serve as the most common examples of industrial chemical waste. However, certain household refrigerants, batteries and cleaning products qualify as chemical waste, too. Depending on the potency of certain chemicals and solvents, as well as the potential safety hazards they present, they may fall under the category of hazardous waste.

Disposal

Because of the safety hazards associated with this type of waste, most chemical waste must be disposed of in a special manner. For industrial forms of chemical waste, the disposal process typically involves sealing the waste in securely sealed chemical-resistant drums or barrels, then transporting the safely stored waste to a special landfill. For household forms of chemical waste, many municipal landfills feature areas specifically designated for this type of waste. Like many forms of hazardous waste, chemical waste tends to be toxic, flammable, corrosive or reactive-or some combination thereof-so it's important to take the proper safety precautions when disposing of it.

Examples of Chemical Waste include, but are not limited to:

- Unused and surplus reagent grade chemicals;
- Intermediates and by-products generated from research & educational experiments;
- Batteries;
- Anything contaminated by chemicals;
- Used oil of all types;
- Spent solvents - including water based;
- Mercury containing items;
- Photographic film processing solutions and chemicals;
- Pesticides;
- Non-returnable gas cylinders;
- Non-empty aerosol cans;
- Chemically contaminated sharps;
- Finely divided powders;
- Contaminated syringes, needles, GC syringes, razor blades, pasteur pipettes, pipette tips;
- Equipment and apparatus containing hazardous waste;
- Computer/electronic equipment;
- Toner cartridges;
- Ethylene glycol;
- Paints - both oil and latex;
- Fluorescent light bulbs;
- Light ballasts;
- Preserved specimens;
- Custodial and industrial cleaners;
- Uncured resins(Phenolic, Epoxy, Styrene, etc.);
- Dye and glazes;

- Degreasing solvents;
- Brake/Transmission/Power Steering Fluids.

Hazardous-Waste Management

Hazardous-waste management is the collection, treatment, and disposal of waste material that, when improperly handled, can cause substantial harm to human health and safety or to the environment. Hazardous wastes can take the form of solids, liquids, sludges, or contained gases, and they are generated primarily by chemical production, manufacturing, and other industrial activities. They may cause damage during inadequate storage, transportation, treatment, or disposal operations. Improper hazardous-waste storage or disposal frequently contaminates surface and groundwater supplies. People living in homes built near old and abandoned waste disposal sites may be in a particularly vulnerable position. In an effort to remedy existing problems and to prevent future harm from hazardous wastes, governments closely regulate the practice of hazardous-waste management.

Hazardous-Waste Characteristics

Hazardous wastes are classified on the basis of their biological, chemical, and physical properties. These properties generate materials that are either toxic, reactive, ignitable, corrosive, infectious, or radioactive.

Toxic wastes are poisons, even in very small or trace amounts. They may have acute effects, causing death or violent illness, or they may have chronic effects, slowly causing irreparable harm. Some are carcinogenic, causing cancer after many years of exposure. Others are mutagenic, causing major biological changes in the offspring of exposed humans and wildlife.

Reactive wastes are chemically unstable and react violently with air or water. They cause explosions or form toxic vapours. Ignitable wastes burn at relatively low temperatures and may cause an immediate fire hazard. Corrosive wastes include strong acidic or alkaline substances. They destroy solid material and living tissue upon contact, by chemical reaction.

Infectious wastes include used bandages, hypodermic needles, and other materials from hospitals or biological research facilities. Radioactive wastes emit ionizing energy that can harm living organisms. Because some radioactive materials can persist in the environment for many thousands of years before fully decaying, there is much concern over the control of these wastes.

However, the handling and disposal of radioactive material is not a responsibility of local municipal government.

Green Chemistry

Green chemistry is also known as environmentally benign chemistry, or sustainable chemistry. Perhaps the most widely accepted definition of green chemistry is the one offered by chemists Paul Anastas and John Warner, who defined green chemistry as the design of chemical products and processes that reduce or eliminate the use and generation of hazardous substances.

Milestones

The Pollution Prevention Act of 1990 set the stage for green chemistry: Its focus is the prevention of pollution at the source rather than the treatment of pollutants after they are formed. This goal became a formal objective of the Environmental Protection Agency (EPA) in 1991. Anastas coined the term "green chemistry" the same year. Two of the most prominent and early advocates of green chemistry were Kenneth Hancock of the National Science Foundation (NSF) and Joe Breen, who after twenty years of service at the EPA then became the first director of the Green Chemistry Institute (GCI) during the late 1990s.

Anastas and Warner formulated the twelve principles of green chemistry in 1998. These serve as guidelines for chemists seeking to lower the ecological footprint of the chemicals they produce and the processes by which such chemicals are made.

Starting in 1996, outstanding examples of green chemistry have been recognized in the United Stateseach year by the Presidential Green Chemistry Challenge (PGCC) awards. These are the only awards in chemistry that are bestowed at the presidential level.

The EPA and the American Chemical Society (ACS) have played a major role in promoting research and development, as well as education, in green chemistry. In 2000 the GCI became a partner of the ACS. Chemical societies around the globe have recognized the importance of green chemistry and promote it through journals, conferences, educational activities, and the formation of GCI chapters. There are GCI chapter affiliates all over the world.

Importance to Industry: The Triple Bottom Line

During the 1990s many industries began to earnestly adopt green chemistry and other sustainable practices. Forward-looking companies realized that the practice of green chemistry not only leads to environmental benefits, but also economic and social benefits. The combination of these three benefits is known as the "triple bottom line" and provides strong encouragement for businesses to develop sustainable products and processes.

The following real-world examples of green chemistry represent the accomplishments of several winners of the PGCC awards. They illustrate how green chemistry impacts a wide array of fields including pharmaceuticals, pesticides, polymers, and many others.

The Concept of Atom Economy

When chemists are considering a compound, they are concerned with the chemical, biological, and physical properties of this compound, and the Method by which the compound is prepared or its synthesis. In order to focus greater attention on waste by-products that are formed during a synthesis, Barry Trost of Stanford University developed the concept of atom economy. This concept deals with the question: How many of the atoms of the reactants are incorporated into the final desired product and how many are wasted by incorporation into by-products? An example of the application of this concept is discussed in the following synthesis of ibuprofen.

Pharmaceuticals: Ibuprofen is the active ingredient in many analgesic and inflammatory drugs such as Advil, Motrin, and Medipren. Beginning in the 1960s, ibuprofen was produced by a six-step synthesis with an atom economy of only 40 percent. This meant that less than half (40 percent) of the weight of all the atoms of the reactants were incorporated in the ibuprofen, and 60 percent were wasted in the formation of unwanted by-products. The annual production of approximately 30 million pounds of ibuprofen by this method resulted in over 40 million pounds of waste. But during the 1990s, the BHC Company developed a new synthesis of ibuprofen with an atom economy of 77 to 99 percent. This synthesis not only produces much less waste, it is also only a three-step process. A pharmaceutical company can thus produce more ibuprofen in less time and with less energy, which results in increased profits.

Pesticides: Dichlorodiphenyl trichloroethane (DDT) is one of the most well-known insecticides. During World War II it saved thousands of Allied lives by killing disease-carrying insects, but during the 1960s the significant environmental damage caused by DDT was brought to the public's attention by Rachel Carson in Silent Spring (1962). As a result of the controversy generated by this book and other media coverage, the substance's use was banned in the United States in 1973.

During the 1960s and 1970s organophosphates largely replaced organo-chlorine pesticides such as DDT. These pesticides rapidly degrade in the environment, but they are much more toxic to mammals. They are deadly to a wide array of insects and kill not only the target organism but also beneficial insects, such as bees and predatory beetles, and can also be harmful to humans.

One approach to producing less environmentally harmful pesticides is to use compounds that destroy only the target organisms. One manufacturer, Rohm & Haas, has developed insecticides that mimic a hormone used only by molting insects. Insects that do not molt are not affected, leaving many beneficial insects unharmed.

A more recent strategy for protecting plants from pests and disease involves the use of genetically altered plants. This method is controversial. Concerns include cross-pollination with unaltered plants and the entry of altered plants into the food supply.

Condensed Principles of Green Chemistry

P - Prevent wastes
R - Renewable materials
O - Omit derivatization steps
D - Degradable chemical products
U - Use safe synthetic methods
C - Catalytic reagents
T - Temperature, Pressure ambient
I - In-Process Monitoring
V - Very few auxiliary substances
E - E-factor, maximise feed in product
L - Low toxicity of chemical products
Y - Yes, it is safe

Green Chemistry's 12 Principles

These principles demonstrate the breadth of the concept of green chemistry:

1. Prevent Waste: Design chemical syntheses to prevent waste. Leave no waste to treat or clean up.

2. Maximize Atom Economy: Design syntheses so that the final product contains the maximum proportion of the starting materials. Waste few or no atoms.

3. Design Less Hazardous Chemical Syntheses: Design syntheses to use and generate substances with little or no toxicity to either humans or the environment.

4. Design Safer Chemicals and Products: Design chemical products that are fully effective yet have little or no toxicity.

5. Use Safer Solvents and Reaction Conditions: Avoid using solvents, separation agents, or other auxiliary chemicals. If you must use these chemicals, use safer ones.

6. Increase Energy Efficiency: Run chemical reactions at room temperature and pressure whenever possible.

7. Use Renewable Feedstocks: Use starting materials (also known as feedstocks) that are renewable rather than depletable. The source of renewable feedstocks is often agricultural products or the wastes of other processes; the source of depletable feedstocks is often fossil fuels (petroleum, natural gas, or coal) or mining operations.

8. Avoid Chemical Derivatives: Avoid using blocking or protecting groups or any temporary modifications if possible. Derivatives use additional reagents and generate waste.

9. Use Catalysts not Stoichiometric Reagents: Minimize waste by using catalytic reactions. Catalysts are effective in small amounts and can carry out a single reaction many times. They are preferable to stoichiometric reagents, which are used in excess and carry out a reaction only once.

10. Design Chemicals and Products to Degrade after Use: Design chemical products to break down to innocuous substances after use so that they do not accumulate in the environment.

11. Analyze In Real Time to Prevent Pollution: Include in-process, real-time monitoring and control during syntheses to minimize or eliminate the formation of byproducts.

12. Minimize the Potential for Accidents: Design chemicals and their physical forms (solid, liquid, or gas) to minimize the potential for chemical accidents including explosions, fires, and releases to the environment.

References

- Hazardous-waste-management, technology: britannica.com, Retrieved 11, January 2020
- Basics-green-chemistry, greenchemistry: epa.gov, Retrieved 23, April 2020
- What-is-chemical-waste: doityourself.com, Retrieved 13, July 2020
- The-Importance-of-Industrial-Waste-Management-227351: johnsrefuse.com, Retrieved 17, February 2020
- Causes-effects-of-industrial-pollution: conserve-energy-future.com, Retrieved 12, June 2020
- Waste-management-and-waste-disposal-methods: conserve-energy-future.com, Retrieved 16, March 2020
- Industrial-waste-types, chemistry: byjus.com, Retrieved 23, May 2020

5

Air Pollution and its Control

The mixture of solid particles and gases within the air is called air pollution. Some of the major types of air pollutants are sulphur oxides, nitrogen oxides, carbon monoxide, carbon dioxide and volatile organic compounds. This chapter discusses in detail these air pollutants as well as green-house effect and climate change.

Biogeochemical Cycles

Carbon Cycle

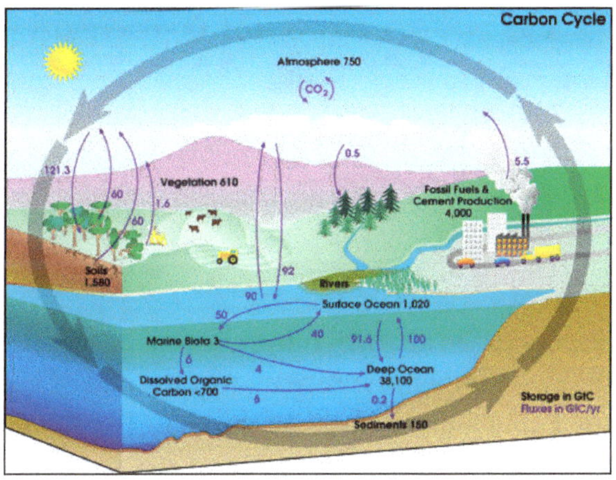

Photosynthesis is a complex series of reactions carried out by algae, phytoplankton, and the leaves in plants, which utilize the energy from the sun. The simplified version of this chemical reaction is to utilize carbon dioxide molecules from the air and water molecules and the energy from the sun to produce a simple sugar such as glucose and oxygen molecules as a byproduct. The simple sugars are then converted into other molecules such as starch, fats, proteins, enzymes, and DNA/RNA i.e. all of the other molecules in living plants. All of the "matter/stuff" of a plant ultimately is produced as a result of this photosynthesis reaction. An important summary statement is that during photosynthesis plants use carbon dioxide and produce oxygen.

Carbon Cycle - Photosynthesis

$$CO_2 + H_2O + \text{energy} \rightarrow C_6H_{12}O_6 + O_2$$

CO_2	H_2O	energy	$C_6H_{12}O_6$	O_2
carbon dioxide	water	sun	glucose represents any organic matter in plants	oxygen

Combustion/Metabolism Reaction

Combustion occurs when any organic material is reacted (burned) in the presence of oxygen to give off the products of carbon dioxide and water and energy. The organic material can be any fossil fuel such as natural gas (methane), oil, or coal. Other organic materials that combust are wood, paper, plastics, and cloth. Organic materials contain at least carbon and hydrogen and may include oxygen. If other elements are present they also ultimately combine with oxygen to form a variety of pollutant molecules such as sulfur oxides and nitrogen oxides.

Metabolism occurs in animals and humans after the ingestion of organic plant or animal foods. In the cells a series of complex reactions occurs with oxygen to convert for example glucose sugar into the products of carbon dioxide and water and energy. This reaction is also carried out by bacteria in the decomposition/decay of waste materials on land and in the water.

An important summary statement is that during combustion/metabolism oxygen is used and carbon dioxide is a product. The whole purpose of both processes is to convert chemical energy into other forms of energy such as heat.

Carbon Cycle - Photosynthesis

$$Ch_4 + O_2 \rightarrow CO_2 + H_2O + \text{energy}$$

or (oxygen) (Carbon dioxide) (water) (Heat)

$C_6H_{12}O_6$

represents any fossil fuels food, organic matter

Sedimentation

Carbon dioxide is slightly soluble and is absorbed into bodies of water such as the ocean and lakes. It is not overly soluble as evidenced by what happens when a can of carbonated soda such as Coke is opened. Some of the dissolved carbon dioxide remains in the water, the warmer the water the less carbon dioxide remains in the water.

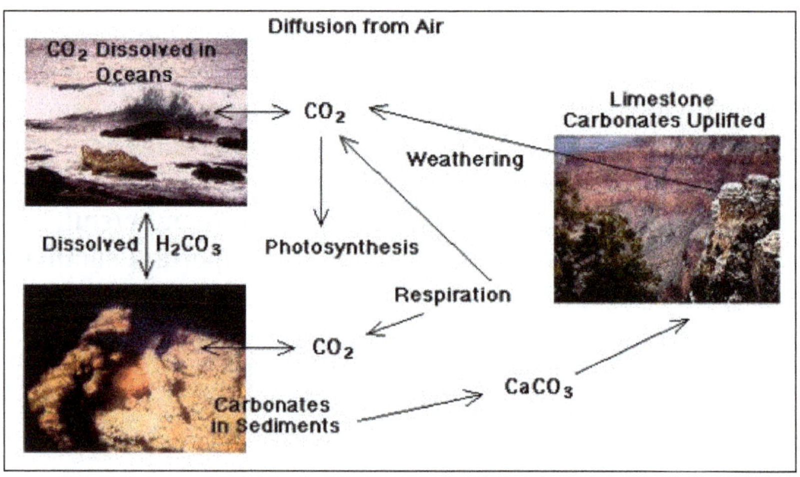

Carbon Cycle-Ocean/Sedimentation.

Some carbon dioxide is used by algae and phytoplankton through the process of photosynthesis.

In other marine ecosystems, some organisms such as coral and those with shells take up carbon dioxide from the water and convert it into calcium carbonate. As the shelled organisms die, bits and pieces of the shells fall to the bottom of the oceans and accumulate as sediments. The carbonate sediments are constantly being formed and re-dissolved in the depths of the oceans. Over long periods of time, the sediments may be raised up as dry land or into mountains. This type of sedimentary rock is called limestone. The carbonates can re-dissolve releasing carbon dioxide back to the air or water.

Human Impacts: Fossil Fuels

In the natural carbon cycle, there are two main processes which occur: photosynthesis and metabolism. During photosynthesis, plants use carbon dioxide and produce oxygen and during metabolism oxygen is used and carbon dioxide is a product. Humans impact the carbon cycle during the combustion of any type of fossil fuel, which may include oil, coal, or natural gas. Fossil Fuels were formed very long ago from plant or animal remains that were buried, compressed, and transformed into oil, coal, or natural gas. The carbon is said to be "fixed" in place and is essentially locked out of the natural carbon cycle. Humans intervene during by burning the fossil fuels. During combustion in the presence of air (oxygen), carbon dioxide and water molecules are released into the atmosphere. The question becomes as to what happens to this extra carbon dioxide that is released into the atmosphere. This is the subject of considerable debate and about it possible effect in enhancing the greenhouse effect which may than result in global warming.

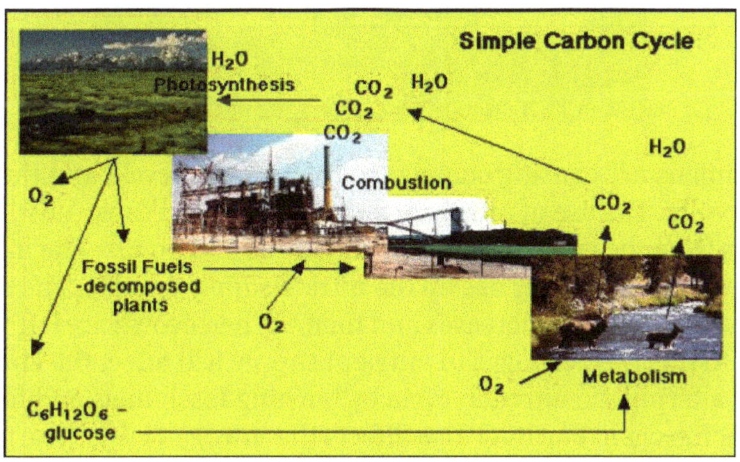

Nitrogen Cycle

Nitrogen is an element that is found in both the living portion of our planet and the inorganic parts of the Earth system. Nitrogen moves slowly through the cycle and is stored in reservoirs such as the atmosphere, living organisms, soils, and oceans along the way.

Most of the nitrogen on Earth is in the atmosphere. Approximately 80% of the molecules in Earth's atmosphere are made of two nitrogen atoms bonded together (N_2). All plants and animals need nitrogen to make amino acids, proteins and DNA, but the nitrogen in the atmosphere is not in a form that they can use. The molecules of nitrogen in the atmosphere can become usable for living things when they are broken apart during lightning strikes or fires, by certain types of bacteria, or by bacteria associated with legume plants. Other plants get the nitrogen they need from the soils or water in which they live mostly in the form of inorganic nitrate (NO_3^-). Nitrogen is a limiting factor for plant growth. Animals get the nitrogen they need by consuming plants or other animals that contain organic molecules composed partially of nitrogen. When organisms die, their bodies decompose bringing the nitrogen into soil on land or into the oceans. As dead plants and animals decompose, nitrogen is converted into inorganic forms such as ammonium salts (NH_4^+) by a process called mineralization. The ammonium salts are absorbed onto clay in the soil and then chemically altered by bacteria into nitrite (NO_2^-) and then nitrate (NO_3^-). Nitrate is the form commonly used by plants. It is easily dissolved in water and leached from the soil system. Dissolved nitrate can be returned to the atmosphere by certain bacteria through a process called denitrification.

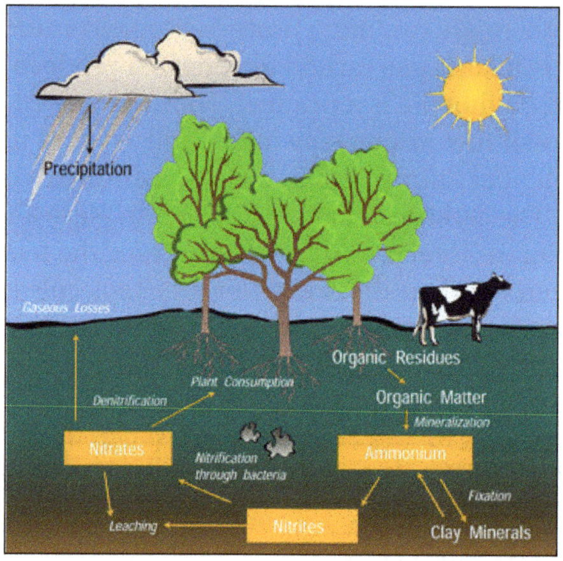

Certain actions of humans are causing changes to the nitrogen cycle and the amount of nitrogen that is stored in reservoirs. The use of nitrogen-rich fertilizers can cause nutrient loading in nearby waterways as nitrates from the fertilizer wash into streams and ponds. The increased nitrate levels cause plants to grow rapidly until they use up the nitrate supply and die. The number of herbivores will increase when the plant supply increases and then the herbivores are left without a food source when the plants die. In this way, changes in nutrient supply will affect the entire food chain. Additionally, humans are altering the nitrogen cycle by burning fossil fuels and forests, which releases various solid forms of nitrogen. Farming also affects the nitrogen cycle. The waste associated with livestock farming releases a large amount of nitrogen into soil and water. In the same way, sewage waste adds nitrogen to soils and water.

Nitrogen and Air Pollution

Nitric oxide (NO) and nitrogen dioxide (NO_2) are together known as nitrogen oxides. These nitrogen oxides contribute to the problem of air pollution, playing roles in the formation of both smog and acid rain. They are released into Earth's atmosphere by both natural and human-generated sources.

Nitric oxide is a colorless, flammable gas with a slight odor. Nitrogen dioxide is a deep red-orange gas that is poisonous but not flammable. It, along with aerosols, is responsible for the reddish-brown color of smog. At high concentrations it is highly toxic, and can cause serious lung damage. Nitrogen dioxide is a strong oxidizing agent, and is thus very reactive with other compounds.

Scientists estimate that between 20 and 90 million tons of nitrogen oxides in produced naturally each year from sources such as volcanoes, oceans, biological decay, and lightning strikes. Human activities add another 24 million tons of nitrogen oxides to our atmosphere annually.

Both NO and NO_2 are formed during high-temperature combustion in the atmosphere, when oxygen combines with nitrogen. The exhaust gases of cars and trucks are major sources of nitrogen oxides, as are the emissions from electrical power generation plants. Automobile exhaust has more NO than NO_2, but once the NO is released into the atmosphere it quickly combines with oxygen in the air to form NO_2.

Nitrogen oxides are at least partially responsible for several types of air pollution. Nitrogen dioxide lends its color to the reddish-brown haze we call smog. Photodissociation of nitrogen dioxide by sunlight produces nitric oxide and ozone in the troposphere, which is another component of smog. A series of chemical reactions transform Volatile Organic Compounds (VOCs) into substances that combine with nitrogen dioxide to produce PAN (Peroxyacytyl nitrate), yet another element in smog. Nitrogen dioxide in the air also reacts with water vapor to form nitric acid, one of the types of acid in acid rain. Nitric oxide concentration in unpolluted air is around 0.01 ppm. In smog, the concentration rises twenty-fold to about 0.2 ppm.

Although nitrogen oxides have gained dubious distinction as pollutants, they are also used beneficially in some industrial processes. Nitric oxide is manufactured on a large scale, and is subsequently used to make nitric acid (HNO_3). To create nitric oxide for industrial uses, chemists combine ammonia (NH_3) with oxygen (O_2), releasing water (H_2O) as a byproduct. Nitrogen compounds derived from nitric acid are used to create chemical fertilizers, explosives, and other useful substances.

Fig: An unsightly haze of smog.

Air Pollutants

Air pollution is a change in the physical, chemical and biological characteristic of air that causes adverse effects on humans and other organisms. The ultimate result is a change in the natural environment and/or ecosystem.

The substances that are responsible for causing air pollution are called air pollution. These air pollutants can be either natural (e.g. wildfires) or synthetic (man-made); they may be in the form of gas, liquid or solid.

Types of Air Pollutants

An air pollutant is known as a substance in the air that can cause harm to humans and the environment. Pollutants can be in the form of solid particles, liquid droplets, or gases. In addition, they may be natural or man-made. Pollutants can be classified as either primary or secondary. Usually, primary pollutants are substances directly emitted from a process, such as ash from a volcanic eruption, the carbon monoxide gas from a motor vehicle exhaust or sulphur dioxide released from factories.

Secondary pollutants are not emitted directly. Rather, they form in the air when primary pollutants react or interact. An important example of a secondary pollutant is ground level ozone — one of the many secondary pollutants that make up photochemical smog.

Major primary pollutants produced by human activity include:

- Sulphur oxides (SOx): SO_2 is produced by volcanoes and in various industrial processes. Since coal and petroleum often contain sulphur compounds, their combustion generates sulphur dioxide. Further oxidation of SO_2, usually in the presence of a catalyst such as NO_2, forms H_2SO_4, and thus acid rain. This is one of the causes for concern over the environmental impact of the use of these fuels as power sources.

- Nitrogen oxides (NOx): Especially nitrogen dioxide is emitted from high temperature combustion. Nitrogen dioxide is the chemical compound with the formula No_2. It is responsible for photochemical smog, acid rain etc.

- Carbon monoxide: It is a colorless, odorless, non-irritating but very poisonous gas. It is a product by incomplete combustion of fuel such as natural gas, coal or wood. Vehicular exhaust is a major source of carbon monoxide.
- Carbon dioxide (CO_2): A greenhouse gas emitted from combustion but is also a gas vital to living organisms. It is a natural gas in the atmosphere.
- Volatile organic compounds: VOCs are an important outdoor air pollutant. In this field they are often divided into the separate categories of methane (CH_4) and non-methane (NMVOCs). Methane is an extremely efficient greenhouse gas which contributes to enhance global warming.

  Other hydrocarbon VOCs are also significant greenhouse gases via their role in creating ozone and in prolonging the life of methane in the atmosphere, although the effect varies depending on local air quality. Within the NMVOCs, the aromatic compounds benzene, toluene and xylene are suspected carcinogens and may lead to leukaemia through prolonged exposure. 1, 3-butadiene is another dangerous compound which is often associated with industrial uses.
- Particulate matter: Particulates, alternatively referred to as particulate matter (PM) or fine particles, are tiny particles of solid or liquid suspended in a gas. In contrast, aerosol refers to particles and the gas together. Sources of particulate matter can be manmade or natural.

  Some particulates occur naturally, originating from volcanoes, dust storms, forest and grassland fires, living vegetation, and sea spray. Human activities, such as the burning of fossil fuels in vehicles, power plants and various industrial processes also generate significant amounts of aerosols.

  Averaged over the globe, anthropogenic aerosols—those made by human activities—currently account for about 10 per cent of the total amount of aerosols in our atmosphere. Increased levels of fine particles in the air are linked to health hazards such as heart disease, altered lung function and lung cancer.
- Persistent free radicals – connected to airborne fine particles could cause cardiopulmonary disease.
- Toxic metals – such as lead, cadmium and copper.
- Chlorofluorocarbons (CFCs) harmful to the ozone layer emitted from products currently banned from use.
- Ammonia (NH_3) emitted from agricultural processes. Ammonia is a compound with the formula NH_3. It is normally encountered as a gas with a characteristic pungent odor. Ammonia contributes significantly to the nutritional needs of terrestrial organisms by serving as a precursor to foodstuffs and fertilizers. Ammonia, either directly or indirectly, is also a building block for the synthesis of many pharmaceuticals. Although in wide use, ammonia is both caustic and hazardous.
- Odors such as from garbage, sewage, and industrial processes.

- Radioactive pollutants produced by nuclear explosions, war explosives, and natural processes such as the radioactive decay of radon.

Secondary pollutants include:

- Particulate matter formed from gaseous primary pollutants and compounds in photochemical smog. Smog is a kind of air pollution; the word "smog" is a portmanteau of smoke and fog. Classic smog results from large amounts of coal burning in an area caused by a mixture of smoke and sulphur dioxide. Modern smog does not usually come from coal but from vehicular and industrial emissions that are acted on in the atmosphere by sunlight to form secondary pollutants that also combine with the primary emissions to form photochemical smog.
- Ground level ozone (O_3) formed from NOx and VOCs. Ozone (O_3) is a key constituent of the troposphere (it is also an important constituent of certain regions of the stratosphere commonly known as the Ozone layer). Photochemical and chemical reactions involving it drive many of the chemical processes that occur in the atmosphere by day and by night. At abnormally high concentrations brought about by human activities (largely the combustion of fossil fuel), it is a pollutant, and a constituent of smog.
- Peroxyacetyl nitrate (PAN) – similarly formed from NOx and VOCs.

Sources of Air Pollution

Sources of air pollution refer to the various locations, activities or factors which are responsible for the releasing of pollutants in the atmosphere. These sources can be classified into two major categories which are:

Anthropogenic Sources

Anthropogenic sources (human activity) mostly related to burning different kinds of fuel:

- "Stationary Sources" include smoke stacks of power plants, manufacturing facilities (factories) and waste incinerators, as well as furnaces and other types of fuel-burning heating devices.
- "Mobile Sources" include motor vehicles, marine vessels, aircraft and the effect of sound etc.
- Chemicals, dust and controlled burn practices in agriculture and forestry management. Controlled or prescribed burning is a technique sometimes used in forest management, farming, prairie restoration or greenhouse gas abatement. Fire is a natural part of both forest and grassland ecology and controlled fire can be a tool for foresters. Controlled burning stimulates the germination of some desirable forest trees, thus renewing the forest.
- Fumes from paint, hair spray, varnish, aerosol sprays and other solvents.
- Waste deposition in landfills, which generate methane. Methane is not toxic; however, it is highly flammable and may form explosive mixtures with air. Methane is also asphyxiate and may displace oxygen in an enclosed space. Asphyxia or suffocation may result if the oxygen concentration is reduced to below 19.5% by displacement.
- Military, such as nuclear weapons, toxic gases, germ warfare and rocketry.

Natural Sources

- Dust from natural sources, usually large areas of land with little or no vegetation.
- Methane, emitted by the digestion of food by animals, for example cattle.
- Radon gas from radioactive decay within the Earth's crust. Radon is a colorless, odorless, naturally occurring, radioactive noble gas that is formed from the decay of radium. It is considered to be a health hazard. Radon gas from natural sources can accumulate in buildings, especially in confined areas such as the basement and it is the second most frequent cause of lung cancer, after cigarette smoking.
- Smoke and carbon monoxide from wildfires.
- Volcanic activity, which produce sulphur, chlorine, and ash particulates.

Effects of Air Pollutants

There are various harmful effects of the air pollutants:

- Carbon monoxide (source-Automobile exhaust, photochemical reactions in the atmosphere, biological oxidation by marine organisms, etc.): Affects the respiratory activity as haemoglobin has more affinity for CO than for oxygen. Thus, CO combines with HB and thus reduces the oxygen-carrying capacity of blood. This results in blurred vision, headache, unconsciousness and death due to asphyxiation (lack of oxygen).
- Carbon di oxide (source-Carbon burning of fossil fuels, depletion of forests (that remove excess carbon dioxide and help in maintaining the oxygen-carbon dioxide ratio): Causes global warming.
- Sulphur dioxide (source-Industries, burning of fossil fuels, forest fires, electric generation plants, smelting plants, industrial boilers, petroleum refineries and volcanic eruptions): Respiratory problems, severe headache, reduced productivity of plants, yellowing and reduced storage time for paper, yellowing and damage to limestone and marble, damage to leather, increased rate of corrosion of iron, steel, zinc and aluminium.
- Hydrocarbons Poly-nuclear Aromatic Compounds(PAC) and Poly-nuclear Aromatic Hydrocarbons(PAH) (source-Automobile exhaust and industries, leaking fuel tanks, leaching from toxic waste dumping sites and coal tar lining of some water supply pipes): Carcinogenic (may cause leukaemia).
- Chloro-fluoro carbons (CFCs) (source-Refrigerators, air conditioners, foam shaving cream, spray cans and cleaning solvents): Destroy ozone layer which then permits harmful UV rays to enter the atmosphere. The ozone layer protects the earth from the ultraviolet rays sent down by the sun. If the ozone layer is depleted by human action, the effects on the planet could be catastrophic.
- Nitrogen Oxides (source-Automobile exhausts, burning of fossil fuels, forest fires, electric generation plants, smelting plants, industrial boilers, petroleum refineries and volcanic eruptions): Forms photochemical smog, at higher concentrations causes leaf damage or affects the photosynthetic activities of plants and causes respiratory problems in mammals.

- Particulate matter Lead halides (lead pollution) (source- Combustion of leaded gasoline products): Toxic effect in man.
- Asbestos particles (source-Mining activities): Asbestosis – a cancerous disease of the lungs.
- Silicon dioxide (source- Stone cutting, pottery, glass manufacturing and cement industries): Silicosis, a cancerous disease.
- Mercury (source- combustion of fossil fuel & plants)-brain & kidney damage.

Air pollutants affect plants by entering through stomata (leaf pores through which gases diffuse), destroy chlorophyll and affect photosynthesis. During the day time the stomata are wide open to facilitate photosynthesis. Air pollutants during day time affect plants by entering the leaf through these stomata more than night.

Pollutants also erode waxy coating of the leaves called cuticle. Cuticle prevents excessive water loss and damage from diseases, pests, drought and frost. Damage to leaf structure causes necrosis (dead areas of leaf), chlorosis (loss or reduction of chlorophyll causing yellowing of leaf) or epinasty (downward curling of leaf), and abscission (dropping of leaves).

Particulates deposited on leaves can form encrustations and plug the stomata and also reduce the availability of sunlight. The damage can result in death of the plant.So2 causes bleaching of leaves, chlorosis, injury and necrosis of leaves. NO_2 results in increased abscission and suppressed growth. O_3 causes flecks on leaf surface, premature aging, necrosis and bleaching.

Peroxyacetyl nitrate (PAN) causes silvering of lower surface of leaf, damage to young and more sensitive leaves and suppressed growth. Fluorides cause necrosis of leaf-tip while ethylene results in epinasty, leaf abscission and dropping of flowers.

Acid Rain

Acid rain is a result of air pollution. When any type of fuel is burnt, lots of different chemicals are produced. The smoke that comes from a fire or the fumes that come out of a car exhaust don't just contain the sooty grey particles that you can see - they also contains lots of invisible gases that can be even more harmful to our environment.

Power stations, factories and cars all burn fuels and therefore they all produce polluting gases. Some of these gases (especially nitrogen oxides and sulphur dioxide) react with the tiny droplets of water in clouds to form sulphuric and nitric acids. The rain from these clouds then falls as very weak acid - which is why it is known as "acid rain".

Acidity is measured using a scale called the pH scale. This scale goes from 0 to 14. 0 is the most acidic and 14 is the most alkaline (opposite of acidic). Something with a pH value of 7, we call neutral, this means that it is neither acidic nor alkaline.

Very strong acids will burn if they touch your skin and can even destroy metals. Acid rain is much, much weaker than this, never acidic enough to burn your skin.

Rain is always slightly acidic because it mixes with naturally occurring oxides in the air. Unpolluted rain would have a pH value of between 5 and 6. When the air becomes more polluted with nitrogen oxides and sulphur dioxide the acidity can increase to a pH value of 4. Some rain has even been recorded as being pH 2.

Vinegar has a pH value of 2.2 and lemon juice has a value of pH2.3. Even the strongest recorded acid rain is only about as acidic as lemon juice or vinegar and we know that these don't harm us - so why do we worry about acid rain?

Acid rain can be carried great distances in the atmosphere, not just between countries but also from continent to continent. The acid can also take the form of snow, mists and dry dusts. The rain sometimes falls many miles from the source of pollution but wherever it falls it can have a serious effect on soil, trees, buildings and water.

Forests all over the world are dying, fish are dying. In Scandinavia there are dead lakes, which are crystal clear and contain no living creatures or plant life. Many of Britain's freshwater fish are threatened; there have been reports of deformed fish being hatched. This leads to fish-eating birds and animals being affected also. Is acid rain responsible for all this? Scientists have been doing a lot of research into how acid rain affects the environment.

It is thought that acid rain can cause trees to grow more slowly or even to die but scientists have found that it is not the only cause. The same amount of acid rain seems to have more effect in some areas than it does in others.

As acid rain falls on a forest it trickles through the leaves of the trees and runs down into the soil below. Some of it finds its way into streams and then on into rivers and lakes. Some types of soil can help to neutralize the acid - they have what is called a "buffering capacity".

Other soils are already slightly acidic and these are particularly susceptible to the effects of acid rain.

Acid rain can affect trees in several different ways, it may:

- Dissolve and wash away the nutrients and minerals in the soil which help the trees to grow.
- Cause the release of harmful substances such as aluminium into the soil.
- Wear away the waxy protective coating of leaves, damaging them and preventing them from being able to photosynthesize properly.

A combination of these effects weakens the trees which mean that they can be more easily attacked by diseases and insects or injured by bad weather. It is not just trees that are affected by acid rain, other plants may also suffer.

It is in aquatic habitats that the effects of acid rain are most obvious. Acid rain runs off the land and ends up in streams, lakes and marshes - the rain also falls directly on these areas.

As the acidity of a lake increases, the water becomes clearer and the numbers of fish and other water animals decline. Some species of plant and animal are better able to survive in acidic water than others. Freshwater shrimps, snails, mussels are the most quickly affected by acidification followed by fish such as minnows, salmon and roach. The roe and fry (eggs and young) of the fish are the worst affected, the acidity of the water can cause deformity in young fish and can prevent eggs from hatching properly.

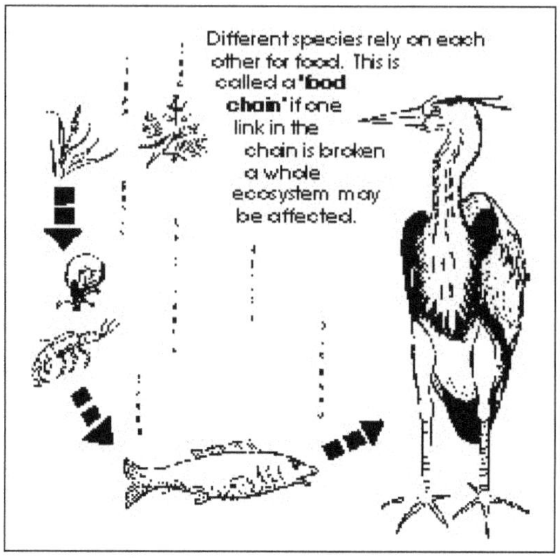

The acidity of the water does not just affect species directly; it also causes toxic substances like aluminium to be released into the water from the soil, harming fish and other aquatic animals.

Lakes, rivers and marshes each have their own fragile ecosystem with many different species of plants and animals all depending on one another to survive. If a species of fish disappears, the animals which feed on it will gradually disappear too. If the extinct fish used to feed on a particular species of large insect, that insect population will start to grow, this in turn will affect the smaller insects or plankton on which the larger insect feeds.

Every type of material will become eroded sooner or later by the effects of the climate. Water, wind, ice and snow all help in the erosion process but unfortunately; acid rain can help to make this natural process even quicker. Statues, buildings, vehicles, pipes and cables can all suffer. The worst affected are things made from limestone or sandstone as these types of rock are particularly susceptible and can be affected by air pollution in gaseous form as well as by acid rain.

Chemistry of Acid Rain

Pure water has a pH of 7.0 (neutral); however, natural, unpolluted rainwater actually has a pH of about 5.6 (acidic). The acidity of rainwater comes from the natural presence of three substances (CO_2, NO, and SO_2) found in the troposphere (the lowest layer of the atmosphere). As is seen in table I, carbon dioxide (CO_2) is present in the greatest concentration and therefore contributes the most to the natural acidity of rainwater.

Gas	Natural Sources	Concentration
Carbon dioxide CO_2	Decomposition	355 ppm
Nitric oxide NO	Electric discharge	0.01 ppm
Sulfur dioxide SO_2	Volcanic gases	0-0.01 ppm

Carbon dioxide, produced in the decomposition of organic material, is the primary source of acidity in unpolluted rainwater.

Parts per million (ppm) is a common concentration measure used in environmental chemistry. The formula for ppm is given by:

$$ppm = \frac{\text{mass of component}}{\text{total mass}} \times 10^6$$

Carbon dioxide reacts with water to form carbonic acid. Carbonic acid then dissociates to give the hydrogen ion (H^+) and the hydrogen carbonate ion (HCO_3^-). The ability of H_2CO_3 to deliver H^+ is what classifies this molecule as an acid, thus lowering the pH of a solution.

$$CO_2 + H_2O \rightarrow H_2CO_3$$

$$H_2CO_3 \rightarrow H^+ + HCO_3^-$$

Nitric oxide (NO), which also contributes to the natural acidity of rainwater, is formed during lightning storms by the reaction of nitrogen and oxygen, two common atmospheric gases. In air, NO is oxidized to nitrogen dioxide (NO_2), which in turn reacts with water to give nitric acid (HNO_3). This acid dissociates in water to yield hydrogen ions and nitrate ions (NO_3^-) in a reaction analogous to the dissociation of carbonic acid shown in Equation below, again lowering the pH of the solution.

$$N_2(g) + O_2 \xrightarrow{\text{lightning}} 2NO(g)$$

$$NO(g) + \frac{1}{2}O_2 \rightarrow NO_2(g)$$

$$3NO_2(g) + H_2O \rightarrow 2HNO_3(aq) + NO(g)$$

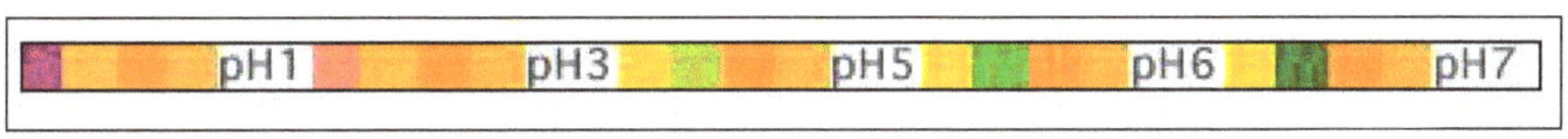

Acidity of Polluted Rainwater

Unfortunately, human industrial activity produces additional acid-forming compounds in far greater quantities than the natural sources of acidity. In some areas of the United States, the pH of rainwater can be 3.0 or lower, approximately 1000 times more acidic than normal rainwater. In 1982, the pH of a fog on the West Coast of the United States was measured at 1.8! When rainwater is too acidic, it can cause problems ranging from killing freshwater fish and damaging crops, to eroding buildings and monuments.

Sources of Excess Acidity in Rainwater

What causes such a dramatic increase in the acidity of rain relative to pure water? The answer lies within the concentrations of nitric oxide and sulfur dioxide in polluted air. As shown in below table and figure, the concentrations of these oxides are much higher than in clean air.

Gas	Non-Natural Sources	Concentration
Nitric oxide NO	Internal Combustion	0.2 ppm
Sulfur dioxide SO_2	Fossil-fuel Combustion	0.1 - 2.0 ppm

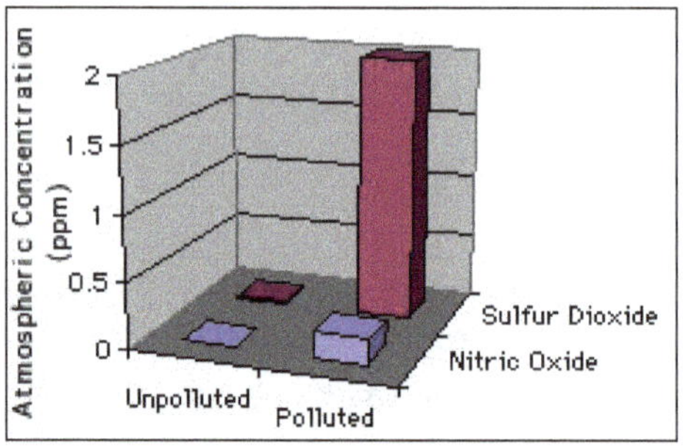

About one-fourth of the acidity of rain is accounted for by nitric acid (HNO_3). In addition to the natural processes that form small amounts of nitric acid in rainwater, high-temperature air combustion, such as occurs in car engines and power plants, and produces large amounts of NO gas. This gas then forms nitric acid. Thus, a process that occurs naturally at levels tolerable by the environment can harm the environment when human activity causes the process (e.g., formation of nitric acid) to occur to a much greater extent.

What about the other 75% of the acidity of rain? Most is accounted for by the presence of sulfuric acid (H_2SO_4) in rainwater. Although sulfuric acid may be produced naturally in small quantities from biological decay and volcanic activity, it is produced almost entirely by human activity, especially the combustion of sulfur-containing fossil fuels in power plants. When these fossil fuels are burned, the sulfur contained in them reacts with oxygen from the air to form sulfur dioxide (SO_2). Combustion of fossil fuels accounts for approximately 80% of the total atmospheric SO_2 in the United States. The effects of burning fossil fuels can be dramatic: in contrast to the unpolluted atmospheric SO_2 concentration of 0 to 0.01 ppm, polluted urban air can contain 0.1 to 2 ppm SO_2, or up to 200 times more SO_2! Sulfur dioxide, like the oxides of carbon and nitrogen, reacts with water to form sulfuric acid.

$$SO_{2(g)} \xrightarrow{O_2} SO_{3(g)} \xrightarrow{H_2O} + H_2SO_4$$

Sulfuric acid is a strong acid, so it readily dissociates in water, to give an H^+ ion and an HSO_4^- ion. The HSO_4^- ion may further dissociate to give H^+ and SO_4^{2-}. Thus, the presence of H_2SO_4 causes the concentration of H^+ ions to increase dramatically, and so the pH of the rainwater drops to harmful levels.

$$H_2SO_4 \rightarrow HSO_4^- + H^+$$

$$H_2SO_4^- \rightarrow SO_4^{2-} + H^+$$

Smog

Smog is a specific type of air pollution. It is a combination of harmful pollutants (often appearing relatively low to the ground as a yellow-brown haze) that are introduced into the atmosphere by both natural and human induced processes. It was first described over 5 decades ago as a mixture of smoke and fog, hence the name "smog"—but today it has a more specific definition and composition.

Smog is made up of many chemicals including nitrogen oxides (NOx), sulphur dioxide (SOx), carbon monoxide (CO), and volatile organic compounds (VOCs), but the two main components of smog are particulate matter (PM) and ground-level ozone (O_3).

Normal smog (often called London-type smog) is mainly a product of burning large amounts of high sulfur coal. Whereas photochemical smog is a more modern phenomena commonly produced by vehicle emissions in contact with sunlight—mostly from burning gasoline and diesel. Photochemical smog forms in warm, densely populated cities with many vehicles.

Formation

Smog forms when pollutants are released into the air. The pollutants are formed both naturally and by humans; however, the human-induced pollutants are of most concern due to the magnitude of pollutants produced by the burning and extraction of fossil fuels, which are known to cause extreme health effects. The location of smog formation is also of great concern, especially for human health, as a good portion of it is produced within cities where large portions of the population live.

Fig: The air in China after rainfall (left) vs on a smoggy day (right).

One of the primary constituents, ozone, is created through chemical reactions between sunlight and certain pollutants. The other primary constituent, particulate matter can also form through chemical reactions but is introduced to the atmosphere through other means as well. Wind may disperse these particles among land, thereby decreasing the amount within a given area. In addition, rainfall may wash these pollutants out of the local atmosphere; however, this can result in other unwanted events like acid rain. When smog encounters an inversion layer (caused by warm areas in the upper atmosphere) it can stay over a region for an extended period of time—exposing people to its effects for longer.

Health Effects of Smog

The health effects from smog and its components can be severe, and depend on many variables. Smog is harmful when inhaled, with the severity of it depending on the amount inhaled, the types of pollutants contained in it, as well as the individual's age, weight, activity level, and well-being. However studies show that any exposure to these pollutants is harmful, with extended exposure and higher doses obviously causing the most damage.

Types of effects smog has include:

- Eye, nose and throat irritation.
- Decreased lung function.
- Aggravation of respiratory or heart disease.
- In some cases, death.

This effect can be seen in figure below, with more serious and rare effects at the top, and the more common yet milder effects towards the bottom.

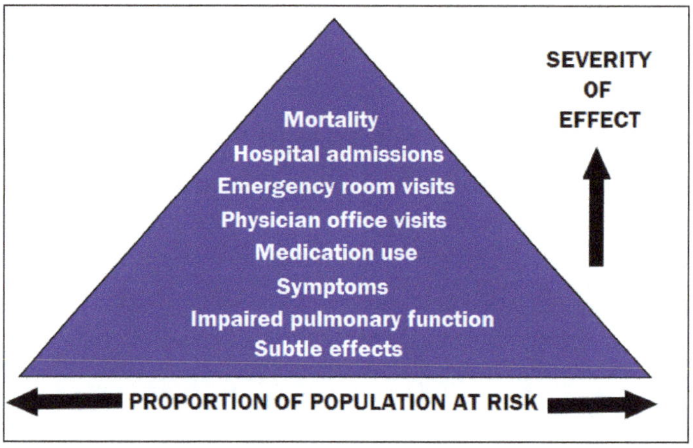

Specific Health Effects of Pollutants

- Particulate matter (PM): Made up of solid and liquid particles that vary in size and composition, most PM gets caught in the nose and throat, and will not reach the lungs. However, smaller particulates can make it to the lungs and are referred to as "respirable PM". These small particles can penetrate deep and cause damage. The cardio-respiratory (heart-lung) system are damaged the most from PM, resulting in infections, asthma, bronchitis, emphysema, black lung and more.
- Ozone (O_3): An intense irritant and dangerous health threat, ozone can damage the eyes and nose, along with the throat and lungs. It is responsible for respiratory problems such as coughing and wheezing, and can reduce lung function. Although irritated cells are replaced a few days after exposure, the replacement is not perfect and constant exposure may lead to aging of the lung and long-term effects.
- Nitrogen oxides (NOx): Responsible both directly and indirectly for health effects, NOx irritates lungs and lowers resistance to infection. Even short-term exposure can cause symptoms of coughing and shortness of breath. Indirectly it is involved in the formation of O_3 and PM.
- Sulfur dioxide (SOx): SO_2 leads to eye irritation, shortness of breath and impaired lung functioning. It can penetrate deeply into the lungs, and when combined with water converts to sulfuric acid which the body is sensitive to. It is also one of the main contributors to the formation of small PM.

- Carbon monoxide (CO): CO is an air pollutant that effects health and in high concentrations is very deadly. It binds with haemoglobin in the blood, reducing its ability to carry oxygen and thereby starves vital tissues of oxygen. People with weaker immune systems and previous health problems are affected the most by high CO levels.

Environmental Impacts

Smog affects much more than just human lives; it also has a large impact on the surrounding physical environment. Ozone and PM in particular cause damaging effects.

Ozone can damage plant cells and inhibit their growth, reducing the amount of carbon dioxide that they take in during photosynthesis. This results in reduced agricultural yields in all sorts of crops. It can also affect synthetic materials like rubber, cotton and other materials, causing deterioration and even disintegration.

Particulate matter smothers plants limiting their exposure to the sun and inhibiting their ability to take in carbon dioxide, therefore, decreasing its photosynthesis capabilities. The chemical composition of the PM may also have an effect on the plants and their surrounding soil. Particulate matter causes the presence of haze in the air, thus reducing the clarity and colour of what can be seen. Humidity of the surrounding air also plays a role in its hazy effect. According to the U.S. Environmental Protection Agency, visibility is reduced from 144 kilometers to around 30 kilometers in certain parts of the United States.

Photochemical Smog

The other name of photochemical smog is “Los Angeles Smog.” This type of smog happens in urban areas that contain a large number of automobiles—one of the adverse examples of this kind of smog pollution in Delhi, the capital of India. When the winter takes hold of the city, we are introduced to foggy and almost choking mornings. That’s because of how many vehicles are being driven on the road producing harmful emission in such a large quantity, that air can’t dissolve.

Effects of Photochemical Smog

Thus, we have two types of smog, the first one is the classic smog, and the second one is photochemical smog. In addition to this, the effects of photochemical smog are much harmful than the other smog. This is because the components of photochemical smog are oxidizing in nature. Some of the common chemicals that you can find in photochemical smog is nitrogen dioxide and tri-oxygen or ozone. As a result, it leads to corrosion of metal which is present in the open along with corrosion of painted surfaces, rubber and stones also.

When there is a high concentration of photochemical smog present in the atmosphere, you will suffer from chest pain, headaches, along with throat dryness. It can cause multiple respiratory ailments if not treated in due time.

Furthermore, photochemical smog contains PAN, acrolein, and formaldehyde. The burning of fossil fuels is the main reason for photochemical smog. When fossil fuels such as petrol and diesel burn, they leave hydrocarbons, that forms ozone, PAN, other chemicals in the atmosphere.

Causes of Smog

Smog is a result of complex photochemical reactions which involves aggressively volatile organic compounds. It is not a single industry or an automobile company which is responsible for the smog in your city. Several factors need to be blamed. But the worst of worst is automobile exhaust as they are present inside the city and they leave harmful gases at the same level where adults and children breathe. On the other hand, we have power plants that are producing electricity by burning up the coal, factories that are making product by burning up fossil fuels. Even consumer products such as paint that you buy for your home, hairspray, chemical solvents, and your plastic food packaging. So these are few lines of work that are causing smog in your city. So when the text time you see smog in the morning you know whom to blame.

Difference Between Classical Smog and Photochemical Smog

There is not much difference between a cloud of smog and photochemical smog. Both of them are harmful to us and for our elderlies. The smog that you see is just a combination of smoke and fog, and it occurs in winter season most of the time in the cities that are packed with lots of vehicles. Smog could happen if there is no wind blowing in your area, and there's been a high level of industrial activity. Smog takes place far away from its actual source. This kind of smog is of sulphurous type and has another name called "London Smog."

While on the other hand, photochemical smog happens due to ultraviolet radiation that reacts to pollutants that are present in the air. Common pollutants present in the air for photochemical smog are nitrogen oxides and hydrocarbons. Sunlight is a must for photochemical smog to take place.

Greenhouse Effect and Climate Change

Mean Temperature of Earth

Visible light from the Sun carries energy to the planets in our solar system. That sunlight is absorbed by the planet's surface, heating the ground. Any object with a temperature above absolute zero emits electromagnetic radiation. For planets, that outgoing EM radiation takes the form of infrared "light". The planet will continue to warm until the outgoing infrared energy exactly balances the incoming energy from sunlight. Scientists call this balance "thermal equilibrium". With a bit of basic physics we can calculate the temperature at which this state of thermal equilibrium is reached.

Satellites have directly measured the amount of energy arriving at Earth from the Sun as sunlight. Although this value varies slightly over time, it is usually very close to 1,361 watts of power per square meter. To visualize this, imagine lighting a small closet with about 13 or 14 one hundred watt light bulbs. Scientist refers to the amount of incoming energy from sunlight as "insolation". The specific value at Earth of 1,361 W/m^2 is called the "solar constant".

In order to calculate the total amount of energy arriving at Earth, we need to know how much area is being lit. We then multiply the area by the insolation (in units of energy flow per unit area) to find out the total amount of incoming energy.

It turns out that we can simplify our calculation of area by noticing that the amount of light intercepted by our spherical planet is exactly the same as the amount that would be blocked by a flat disk with the same diameter as Earth, as shown in the diagram below.

The area of a circle is pi times the radius of the circle squared. In this case, the circle's radius is simply the radius of Earth, which is about 6,371 km (3,959 miles) on average. If we multiply this area by the amount of energy per unit area - the solar "insolation", we can determine the total amount of energy intercepted by Earth:

$$E_{intercepted} = K_s \times \pi R_E$$

- E = total energy intercepted (technically, energy flux = energy per unit time, in watts).
- K_S = solar insolation ("solar constant") = 1,361 watts per square meter.
- R_E = radius of Earth = 6,371 km = 6,371,000 meters.

Plugging in values and solving for E, we find that our planet intercepts about 174 petawatts of sunlight; quite a lot of energy.

$$E = 1361 \times 3.1416 \times (6{,}371{,}000)^2$$

$$E = 173.5 \times 10^{15} \text{ watts}$$

Since Earth is not completely black, some of this energy is reflected away and not absorbed by our planet. Scientists use the term albedo to describe how much light a planet or surface reflects away. A planet completely covered with snow or ice would have an albedo close to 100%, while a completely dark planet would have an albedo close to zero. To determine how much energy Earth absorbs from sunlight, we must multiply the energy intercepted times one minus the albedo value; since albedo represents the light reflected away, one minus albedo equals the light energy absorbed. Our equation for total energy absorbed becomes:

$$E_{absorbed} = K_S \times (1 - \text{albedo}) \times \pi R_E^{\,2}$$

Now that we have a value for the energy flowing into the Earth system, let's calculate the energy flowing out.

The sunlight Earth absorbs heats our planet. Any object with a temperature above absolute zero emits electromagnetic (EM) radiation. In the case of Earth, this EM radiation takes the form of longwave, infrared radiation (or IR "light").

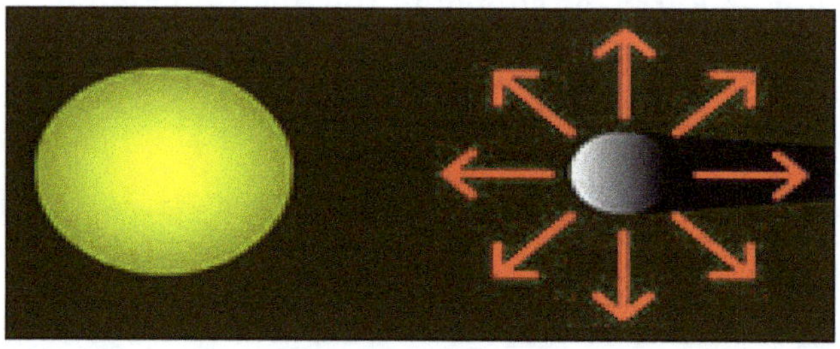

In the 1800s, two scientists determined that the amount of radiation emitted by an object depends on the temperature of the object. The equation for this relationship is called the Stefan-Boltzmann law. It was determined experimentally by Joseph Stefan in 1879 and theoretically derived by Ludwig Boltzmann in 1844. Notice that the amount of energy emitted is proportional to the 4th power of the temperature. Energy emissions go up a lot as temperature rises:

$$j^* = \sigma T^4$$

- j^* = energy flux = energy per unit time per unit area (joules per second per square meter or watts per square meter).
- σ = Stefan-Boltzmann constant = 5.670373×10^{-8} watts / $m^2 K^4$ (m = meters, K = kelvins).
- T = temperature (in the Kelvin scale).

The Stefan-Boltzmann law tells us how much infrared energy Earth will emit per unit area. We need to multiply this by the total area of Earth's surface to calculate the total amount of energy emitted by Earth. Since Earth rotates, its entire surface is heated by sunlight. Therefore, the whole surface of the spherical planet emits infrared radiation. We can't use the same shortcut we used for incoming sunlight by treating Earth as equivalent to a disk. Geometry tells us that the surface area of a sphere is four times pi times the radius of the sphere squared. Multiplying the energy emissions per unit area times the surface area of Earth; we derive an expression for Earth's total infrared energy emissions:

$$E_{emitted} = \sigma T^4 \times 4\pi R_E^{\ 2}$$

The law of conservation of energy tells us that the energy emitted must be equal to the energy absorbed. Setting these two values equal, we can substitute in the expressions for each. Noting that pi times Earth's radius squared appears on both sides of the equation, we can use a little algebra to simplify the result:

$$E_{absorbed} = E_{emitted}.$$

$$K_S \times (1 - \text{albedo}) \times \pi R_E^{\ 2} = \sigma T^4 \times 4\pi R_E^{\ 2}$$

$$K_S \times (1 - \text{albedo}) = 4\sigma T^4$$

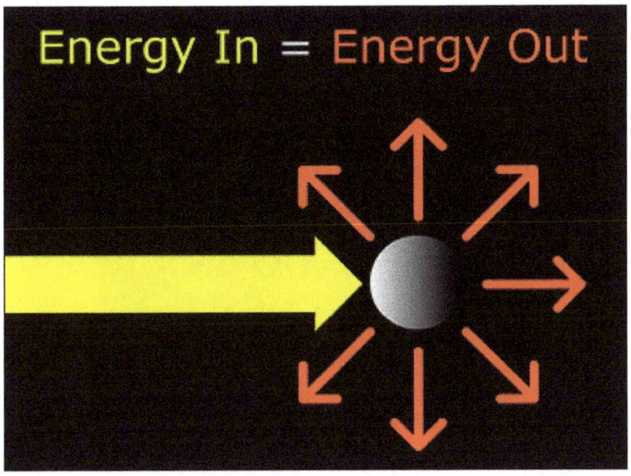

Since the values for the solar constant (K_S), Earth's albedo, and the Stefan-Boltzmann constant (σ) are all known, it is possible to solve this equation for temperature (T). Using a little more algebra, we can write the expression above as:

$$T = \sqrt[4]{\frac{K_S \times (1 - \text{albedo})}{4\sigma}}$$

Earth's overall, average albedo is about 0.31 (or 31%). The value of the Stefan-Boltzmann constant (σ) is 5.6704 x 10^{-8} watts / m^2 K^4. Plugging these numbers and the value for K_S into the equation, we can calculate Earth's expected temperature:

$$T = \sqrt[4]{\frac{1361 \times (1 - 0.31)}{4 \times 5.6704 \times 10^{-8}}} = 253.7\,\text{kelvins}$$

Converting to the more familiar Celsius and Fahrenheit temperature scales, we get:

$$T_{celcius} = T_{kelvin} - 273.15 = -19.5^{\circ}C$$

$$T_{Fahrenheit} = \frac{9}{5} \times T_{celcius} + 32 = -3.1^{\circ}F$$

Based on this calculation, Earth's expected average global temperature is well below the freezing point of water.

Earth's actual average global temperature is around 14°C (57°F). Our planet is warmer than predicted by 34°C (60° F). That's a pretty big difference.

Why is Earth's temperature so much warmer than our calculations predicted? Certain gases in the atmosphere trap some extra heat, warming our planet like a blanket. This extra warming is called the greenhouse effect. Without it, our planet would be a frozen ball of ice. Thanks to the natural greenhouse effect, Earth is comfortable place for life as we know it. However, too much of a good thing can cause problems. In recent decades, a rise in the amount of greenhouse gases has begun to warm Earth a bit too much.

This calculation of the expected temperature can be done for other planets as well. To do so, you need to adjust the value of the solar insolation, K_S. A planet closer to the Sun receives more energy, so K_S is larger. Planets further from the Sun than Earth receive less sunlight, so K_S has a smaller value. Knowing the planet's distance from the Sun, you can make a ratio with Earth's distance and determine the solar insolation at that planet's distance. Since the amount of sunlight decreases as the square of the distance from the Sun, a planet twice as far from the Sun as Earth would receive on one-fourth as much solar energy.

This energy balance calculation also helped scientists discover a bit of a puzzle from Earth's history. Based on observations of similar stars, astronomers think our Sun is brighter now than it was early in its lifetime. The early Sun was probably only about 70% as bright as it is in modern times. If you multiply K_S by 0.7 and use the result for the solar insolation of the early Sun, you'll find that Earth would have been much, much colder than it is today. However, there is a lot of geologic evidence that there was liquid - not frozen - water on Earth even very early in our planet's history. How could Earth have liquid water if it was so cold due to the dimmer Sun? This puzzle is called the Faint Young Sun Paradox. This paradox is an area of active scientific research. Some scientists think early Earth may have had much, much higher concentrations of greenhouse gases in its atmosphere; enough to warm the planet above freezing despite the dimmer Sun.

Greenhouse Effect

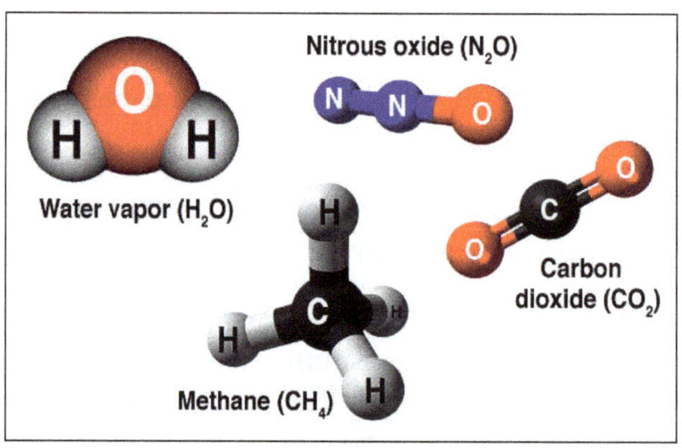

Certain gases in the atmosphere block heat from escaping. Long-lived gases that remain semi-permanently in the atmosphere and do not respond physically or chemically to changes in temperature are described as "forcing" climate change. Gases, such as water vapor, which respond physically or chemically to changes in temperature, are seen as "feedbacks."

Gases that contribute to the greenhouse effect include:

- Water vapor. The most abundant greenhouse gas, but importantly, it acts as a feedback to the climate. Water vapor increases as the Earth's atmosphere warms, but so does the possibility of clouds and precipitation, making these some of the most important feedback mechanisms to the greenhouse effect.
- Carbon dioxide (CO_2): A minor but very important component of the atmosphere, carbon dioxide is released through natural processes such as respiration and volcano eruptions and through human activities such as deforestation, land use changes, and burning fossil fuels. Humans have increased atmospheric CO_2 concentration by 48% since the Industrial Revolution began. This is the most important long-lived "forcing" of climate change.
- Methane: A hydrocarbon gas produced both through natural sources and human activities, including the decomposition of wastes in landfills, agriculture, and especially rice cultivation, as well as ruminant digestion and manure management associated with domestic livestock. On a molecule-for-molecule basis, methane is a far more active greenhouse gas than carbon dioxide, but also one which is much less abundant in the atmosphere.
- Nitrous oxide: A powerful greenhouse gas produced by soil cultivation practices, especially the use of commercial and organic fertilizers, fossil fuel combustion, nitric acid production, and biomass burning.
- Chlorofluorocarbons (CFCs): Synthetic compounds entirely of industrial origin used in a number of applications, but now largely regulated in production and release to the atmosphere by international agreement for their ability to contribute to destruction of the ozone layer. They are also greenhouse gases.

On Earth, human activities are changing the natural greenhouse. Over the last century the burning of fossil fuels like coal and oil has increased the concentration of atmospheric carbon dioxide (CO_2). This happens because the coal or oil burning process combines carbon with oxygen in the air to make CO_2. To a lesser extent, the clearing of land for agriculture, industry, and other human activities has increased concentrations of greenhouse gases.

The consequences of changing the natural atmospheric greenhouse are difficult to predict, but some effects seem likely:

- On average, Earth will become warmer. Some regions may welcome warmer temperatures, but others may not.
- Warmer conditions will probably lead to more evaporation and precipitation overall, but individual regions will vary, some becoming wetter and others dryer.
- A stronger greenhouse effect will warm the ocean and partially melt glaciers and ice sheets, increasing sea level. Ocean water also will expand if it warms, contributing further to sea level rise.

- Outside of a greenhouse, higher atmospheric carbon dioxide (CO_2) levels can have both positive and negative effects on crop yields. Some laboratory experiments suggest that elevated CO_2 levels can increase plant growth. However, other factors, such as changing temperatures, ozone, and water and nutrient constraints, may more than counteract any potential increase in yield. If optimal temperature ranges for some crops are exceeded, earlier possible gains in yield may be reduced or reversed altogether.

Climate extremes, such as droughts, floods and extreme temperatures, can lead to crop losses and threaten the livelihoods of agricultural producers and the food security of communities worldwide. Depending on the crop and ecosystem, weeds, pests, and fungi can also thrive under warmer temperatures, wetter climates, and increased CO_2 levels, and climate change will likely increase weeds and pests.

Finally, although rising CO_2 can stimulate plant growth, research has shown that it can also reduce the nutritional value of most food crops by reducing the concentrations of protein and essential minerals in most plant species. Climate change can cause new patterns of pests and diseases to emerge, affecting plants, animals and humans, and posing new risks for food security, food safety and human health.

- Not enough greenhouse effect: The planet Mars has a very thin atmosphere, nearly all carbon dioxide. Because of the low atmospheric pressure, and with little to no methane or water vapor to reinforce the weak greenhouse effect, Mars has a largely frozen surface that shows no evidence of life.

- Too much greenhouse effect: The atmosphere of Venus, like Mars, is nearly all carbon dioxide. But Venus has about 154,000 times as much carbon dioxide in its atmosphere as Earth (and about 19,000 times as much as Mars does), producing a runaway greenhouse effect and a surface temperature hot enough to melt lead.

Air Pollution Control

These pollutants include fine particulates, carbon monoxide, sulfur dioxide, nitrogen dioxide, ozone etc. Most of the pollutants are organic in nature. The efforts should be taken to control these pollutants. Best method to protect the air quality is to reduce the release of pollutants by changing or improving industrial processes. Pollution control equipment is used in techniques of controlling air pollution. Control equipment removes the pollutants and converts it to less harmful contaminant that results in the recovery of a valuable material for further use.

Removal of Suspended Particulates

Particles can be removed from air by a variety of physical or chemical processes. Some common types of equipment are cyclones, scrubbers, electrostatic precipitators, and baghouse filters. The collected particulates can be removes from the equipment and disposed in a landfill. The important characteristics of particulates on which the selection of the collection devices depends include corrosively, reactivity, shape, density, and especially size and size distribution.

Cyclone

The cyclone collectors are used to control industrial dust emissions. They use centrifugal forces for removing the fine particles. They are also called centrifugal or inertial separators. It contains vertically placed cylinder with an inverted cone attached to base. When the particulate enters tangentially inside it the centrifugal forces tend to drive the particulates to the walls of the cyclone.

The vortex of the cone inside it moves upward after reaching at the bottom of the cylinder. The clear gas is released out from the top of the cylinder on the other hand dust particles are stored at the bottom by gravity. The efficiency of a cyclone mainly depends upon the cyclone diameter. Its efficiency is inversely proportional to the diameter therefore, Smaller the diameter, greater is the efficiency. This is due to increase of centrifugal action with decreasing radius of rotation. The followings are the advantages of cyclone method for removal of particulates:

- It has low initial cost for controlling industrial emissions.
- It is simple in operation as well as construction and for construction it uses those materials which can withstand the temperature and pressure requirements.
- Also it requires low cost for further maintenance.
- The pressure drop of the cyclone used for the removal of particulates is low.
- It is involved in continuous disposal of solid particulate matter.

Electrostatic Precipitators

Electrostatic precipitators are particulate collection devices that use electrostatic force to remove the small particles. The electrostatic precipitators are generally fitted to factories and power station chimneys to collect dust that would otherwise escape into the atmosphere as pollution. It consists of oppositely charge electrodes. The maintenance is required for a high potential difference between the two electrodes. Out of these two electrodes one is a discharging electrode and the other is a collecting electrode. The high potential difference between the electrodes causes a high ionizing field to form.

The gases coming out of the factories pass through negatively charged electrode becoming negatively charged. If the gases passed over positively charged electrodes they get attached to the chimney wall. When they make contact with the electrodes, they become neutralized and may fall into dust collecting space. The collecting spaces are then shaken to further remove dust particles. As a result the gas that enters the atmosphere is free of dust.

Scrubbers

Wet scrubbers are usually used in the case of control of flammable or explosive dusts released from industrial and chemical processes. Generally water is used as the scrubbing fluid. In this process the dust particles comes in contact with water droplets and collide with them. Particulate matter is removed by diffusion of the particulate matter on the liquid medium.

And the condensation of the liquid medium on the particulate matter helps in increasing the size and weight of the particles. And make them easy to remove. The following are the applications of the scrubbers:

- They are particularly used in the case in which the particulate matter is combustible or if any flammable gas is present.
- Scrubbers can be used when there are waste water treatment systems available on the site.
- They are used in the situation when gas reaction and absorption both are simultaneously required with particulate control.

Fabric- Filter Baghouse

Fabric filtration is one of the most common techniques to collect particulate matter from industrial waste gases. It is based on the principle of filtration, which is a reliable, efficient and economic method to remove particulate matter from the gases. The air pollution control equipment using fabric filters are known as bag houses. A bag house consists of numerous vertically hanging, tubular bags which are suspended with their open ends. The number of bags can vary from hundreds to thousands depending upon the size of the bag house. The bags are housed in a shell made of rigid metal material. Hoppers are used to store the collected dust temporarily before it is disposed in a landfill. A discharge device is used for emptying the hopper. The gas entering the pipe strikes a plate, which causes larger particles to fall into a hopper due to gravity. The carrier gas then flows upward into the tubes and outward through the fabric leaving the particulate matter insides the bags.

Control of Toxic Gases

Control of Oxides of Sulfur

Sulfur oxides have many undesirable effects on environment. There are many methods to reduce the amount of sulfur oxides released to the atmosphere. Sulfur in organic compounds can be converted into number of compounds by oxidation or reduction for example sulfur can be oxidize to Sulfur Dioxide and further to Sulfur Trioxide. In the atmosphere, SO_3 reacts with moisture to give sulfuric acid. The most widely used method for control of SOX is the flue-gas desulfurization techniques.

The Flue-Gas and Fuel Desulfurization Techniques

In this process the fuel is burnt in such a way that the sulfur oxides are not emitted to the atmosphere. The process used for this is the catalytic reduction of sulfur dioxide. Sulfur dioxide can be directly converted into elemental sulfur when the gas reacts with carbon monoxide in the presence of a catalyst. Mostly, palladium, silver, cobalt or nickel is used as catalysts.

$$2CO+SO_2 \rightarrow 2CO_2+S$$

Physico-chemical method is also used for controlling sulfur oxides in fuel desulfurization techniques. This is basically used to remove sulfur from coal before combustion takes place. For the removal of sulfur from coal it is first subjected to grinding and then hydraulic washing. During the process inorganic sulfur being heavier tends to settle down and organically bound sulfur in coal is removed by passing H_2S gas in the presence of mixed cobalt-molybdenum oxide catalyst. The organic sulfur is then converted into hydrogen sulfide which is then oxidized to sulfur. Then the sulfur may be used as raw material.

$$H_2S+(O) \rightarrow H_2O+S$$

Control of oxides of Nitrogen Fossil fuel based power plants and automobile exhausts are the main sources of oxides of Nitrogen. The oxide of nitrogen can be controlled by following two methods:

- Flue gas treatment.
- Fluidized bed combustion.

Flue Gas Treatment

Flue gas treatment can be used to control the oxides of nitrogen. In this process, ammonia is added to the flue gas prior to the gas passing over a catalyst. The catalyst enables the ammonia to react chemically with the NOx converting it to molecular nitrogen and water and 90% of the nitrogen oxides remove from the flue gases. Most commonly metallic platinum is used as a catalyst in this reaction.

Fluidized Bed Combustion

This process suppresses the level of sulfur dioxide from the flue gases after combustion. Limestone is added to the fluidized mixture to reduce the amount of sulfur dioxide emitted in the flue gases. The following reactions will take place:

$$CaCO_3 \rightarrow CaO + CO_2$$

$$2CaO + SO_2 + O_2 \rightarrow CaSO_4$$

Noxout Technology

It is a modern technology which uses urea to rid flue gases to nitrogen oxides. It involves the injection of aqueous solution of urea in the combustion system. Urea interacts with nitrogen oxides as a result of which Carbon dioxide, Nitrogen and water is formed. It is estimated that about 80% of the nitrogen oxides are removed by noxious technology.

$$NH_2CONH_2 + NO + NO_2 \rightarrow 2N_2 + CO_2 + 2H_2O$$

Control Measures for Carbon Monoxide

Catalytic burners are used to oxides toxic carbon monoxide into harmless carbon dioxide gas. The toxic gas has potential to hazard in space fights and submarines. Therefore in order to reduce the emission of carbon monoxide from vehicles, the automobile companies us catalytic converter.

Greenhouse Gases

Greenhouse gases are the gases which allow the sun radiation to pass through and reach to earth's surface and results in global warming. It includes carbon dioxide, methane, nitrous oxide, ozone and chlorofluorocarbons. It affects climate, plants, human health and wildlife. For restriction in emissions of greenhouse gases alternative fossil fuels should be used. These energy resources will reduce the emission of harmful oxides of sulfur and nitrogen. Also, the reduction in deforestation by growing more trees is valuable because it would help in preventing drought and improve the soil fertility.

Photochemical Smog

The main constituent of photochemical smog is from automobile exhaust. Photochemical smog has adverse effects on environment. The three components of photochemical smog are nitrogen oxide, ozone and organic derivatives each contributes to the hazardous effects of the smog. The photochemical smog can be suppressed by the following method.

Chemical Method

Diethylhydroxylamine is sprayed to the air and generate free radicals which readily combine with free radical of photochemical smog. It suppressed the formation of smog and freshen-up the polluted air. This will reduce the production of nitrogen dioxide.

Also, catalytic converter can be used in automobiles for reducing smog formation to prevent the release of nitrogen oxide.

Control of Volatile Organic Compounds

Control and treatment of volatile organic compound and organic hazardous air pollutant emissions are done by the following techniques.

Adsorption

This is one of the most commonly used methods, especially for controlling emissions from small sources. It can be physical adsorption or chemisorption. Chemisorption is rarely used for the emission control of volatile organic compounds because it involves a less-reversible chemical bonding of the pollutant and the adsorbing solid.

Physical adsorption uses Van der Waal forces therefore the adsorption process is reversible in nature due to the weaker bonding of the gas and adsorbent material. Activated carbon is a commonly used adsorbent because of its high surface area and material hardness. Then for the regeneration process generally steam desorption is done.

Incineration or Combustion Method

Incineration or combustion is another common technology. Complete combustion or oxidation of pure hydrocarbons produces carbon dioxide and water. Sulfur and nitrogen compounds produce acid gases and limited air supply results in the formation of carbon monoxide. Condensation is used to recover gasoline and fuel vapors at gasoline loading terminals and in gasoline dispensing facilities. It is also used in the adsorbent regeneration process to separate solvents from the stream to regenerate the activated carbon.

Continuous Monitoring Instruments

Continuous monitoring instruments are important for the determination of emission factors in atmosphere. Continuous monitoring basically makes use of analytical methods for the determination of the concentration of air pollutants and their effect on environment. Detailed analysis is also required to determine the emitting rates of air pollutants. The air pollution usually in urban areas is monitored by locating certain number of monitoring stations. The quality of air should be monitored for maintaining health and environment. These instruments are available for the measurement of indoor and outdoor air pollution. It helps people to understand the quality of air around them. These instruments include concentration measurement instruments, continuous emission monitoring systems, air measuring devices and meteorological instruments. These instruments are used to measure gaseous as well as particulate matter.

It can be called as an environmental monitoring which is used to describe the processes and activities that are used to characterize and monitor the quality of the environment. The monitoring instrument also have some limitations as they require high investment costs, good maintenance and highly specialized training. The new approaches are suggested in order to overcome these difficulties especially in those countries where investment and maintenance costs for these instruments are major problems.

The following are the types of continuous monitoring instruments based on used for continuous measurement of indoor or outdoor pollution.

Gasman II

This instrument is used to monitor gases like H_2S, CO, CH_4, or O_2. The lightweight instrument has highly visible and audible alarms. It has data logging capability which is windows compatible PC interface and it is very safe to be used in hazardous areas. Hazardous gas monitor fatly measures even minute concentrations of toxic gases such as chlorine, ammonia, sulfur dioxide and ozone. It has a long sensor life and extremely low cost is required for its maintenance. The calibration of instrument is simple and it is very easy to monitor.

Carbon Dioxide Detector

As the name suggest it is designed to measure the concentration of CO_2. It is used in many industries like agriculture, horticulture, nuclear power plants, food processing industry and the beverage industry. This monitoring instrument is compact, lightweight and provided with a quick and easy calibration.

CEA Gas Monitor

CEA gas monitors are used for sensing ammonia, hydrocarbons, ethylene oxide, hydrogen combustibles, sulfur dioxide, hydrogen sulfide, carbon monoxide and others. The sensors are highly specific, fast responding and resist to poison. They are not affected by moisture or temperature changes.

Control Instrument Corp

The control instruments corp is used to handle readings from up to four different types of sensors. It includes combustible gas, oxygen and toxic gases such as hydrogen sulfide. The device is featured with all alarm and display.

Single-Point Gas Monitor

It is a compact, pocket sized, single gas microprocessor controlled monitor and used to continuously detect and displays gas concentration on an easy to read large digital display. It is required for potential gas leak areas. The features of the monitoring instrument includes that there is a choice of local and remote sensing with this instrument, is has a flashing LEDs facility for warning in any alarming situation, it displays the type of gas and its concentration during detection. It can be monitor up to two gases with local or remote sensors. It is designed to detect oxygen deficiency as well as toxic and combustible gases in air which includes carbon monoxide, hydrogen sulfide, sulfur dioxide, chlorine, and chlorine dioxide. Also in this monitoring instrument a catalytic sensor is used for detecting combustible gases.

Multi-Gas Monitor

The multi-gas monitoring instruments have multi-gas-sensing technology along with an internal sampling pump. This pump is used to provide remote detection facilities in one complete package. They are basically used for industrial purposes which include industries like oil, gas, petrochemical, pulp, paper, food, beverage, public utility, municipal waste and heavy industrial manufacturing. It has many advantageous features first of all the major advantage is as named suggests it is used to monitor multi-gas and other advantages include automatic zero and automatic span calibration, two alarm levels as per sensor and a Peak alarm.

Single-Gas Fixed Monitor

Single gas fixed monitor is a Safety Area Monitor from Unimax. It is used in those area or space where continuous safety verification is needed. It is an effective monitoring instrument for toxic gases like CO, NH_3, H_2S, SO_2, NO, NO_2, HCN, Cl_2. It is also used in determining oxygen levels. Safety area monitor provides maximum safety and protection to workers.

Some other types of monitoring instruments are:

- Emissions Monitoring: This type of monitoring will be focused on emissions coming out of natural and man-made sources like automobiles and industries.
- Ambient Monitoring: This is used for monitoring the ambient air concentration of toxic as well as non-toxic pollutants.
- Deposition Monitoring: This is used to especially measure the dry and wet deposition of air pollutants.
- Visibility Monitoring: This is primarily used to see atmospheric contaminants for monitoring.
- Upper Air Monitoring: This is used to look at ambient concentrations in upper atmosphere with the help of satellites, airplanes etc.

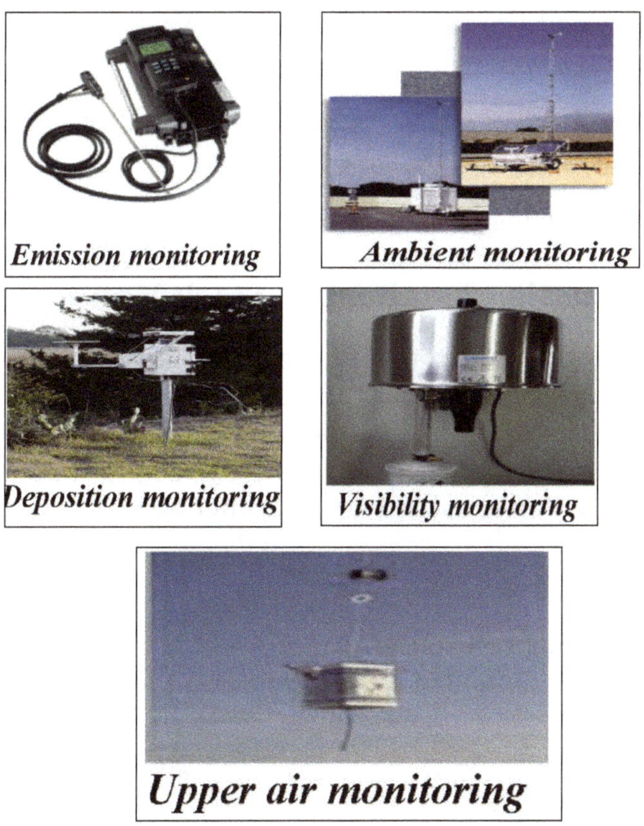

Fig: Types of continuous monitoring instruments.

The choice of monitoring instrument used for analysis will depend upon factors such as cost of the monitoring instrument, the air pollutant and lastly the environment in which the monitoring equipment is required. There are two types of calibration methods used in air monitoring instruments i.e. static methods and dynamic methods. In Static methods a simple one point electrical or chemical test is required. On the other hand dynamic methods are based on generating a flowing stream of calibration gas and then the whole instrument.

Continuous Emission Monitoring Instrument

It is historically used to provide the information in industrial areas about combustion control as the industries are responsible for emission of hazardous gases like carbon monoxide and carbon dioxide. The continuous emission monitoring system is designed by having a sample collector, filter, gas conditioning system calibration gas system and a series of gas analyzers. The monitored emissions by these instruments include sulfur dioxide, nitrogen oxides, carbon monoxide, carbon dioxide, hydrogen chloride, airborne particulate matter, mercury, volatile organic compounds and oxygen. The system can also be used to measure air flow and moisture.

In monitoring these emissions, the system must be in continuously collects, record and report the required emissions data of air pollutants. In the monitoring process a small sample of flue gas is extracted, by a pump and delivered into the continuous emission monitoring system via a sample collector. Dilution is also done because the pure flue gas can be hot, wet and can be stick with any pollutant. The diluted sample is then transported through a sample line to the gas analyzer. The gas analyzers can use various techniques to measure the concentrations of air pollutants. The techniques include infrared and ultraviolet adsorption, chemiluminescence, fluorescence and beta ray absorption.

Continuous emission monitoring system is basically designated to continuously obtain direct emission measurements. This is the most reliable method.

The system includes the following instruments given below.

Unimax Continuous Emission Monitoring- Stack Monitoring

It is a reliable continuous emission monitoring system. It mainly measures low level HCl gas and some other gas pollutants such as SO_2, NOx, CO and CO_2. In its mechanism the dust and moisture from the sample is directly removed directly by the sampling system of the instrument when the sample was get extracted from the stack. The obtained dry sample is then transferred to the gas analyzer in which infrared radiation is utilized with gas filter correlation.

Computerized Stack Sampler

The Computerized Stack Sampling instrument can control the flow of gaseous pollutants so that the sampling is always isokinetic. Sampling can also be performed with a high stack gas low rates and heavy filter loading.

Combustion Gas Analyzer

A combustion gas analyzer collects and stores data independently for up to 48 hours. It measures gases like O_2, CO, NOx, NO, NO_2, SO_2, Rh and F. It also helps in measuring efficiency, draft and pressure. There is a new thick film sensor its performance is superior to the catalytic bead detectors. It includes some additional Features like a menu driven user interface and LCD display. It is also designated with an Auto calibration technique.

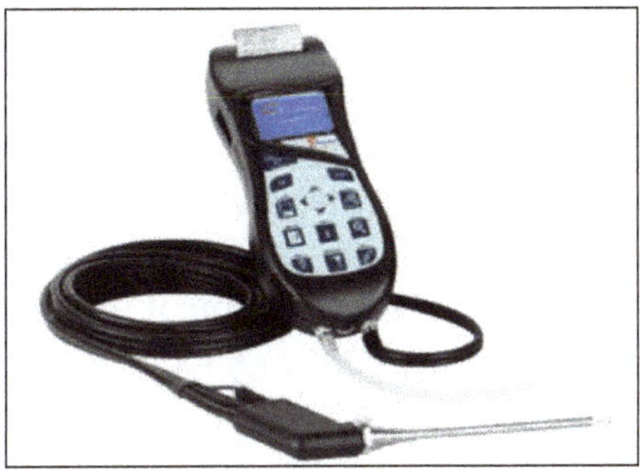

Fig: Portable industrial combustion gas analyzer.

Combustible Gas Transmitter

The new infrared combustible gas transmitter requires less maintenance than catalytic bead devices. It has many other advantages such as there is no risk of positioning in sensor from silicon, halogens, and other contaminants, the high concentration of gases will cause no saturation to sensor also there is no requirement of oxygen for operation. Even if the sensor is continuously exposed to gas there will be no reductions in sensor life as its response time is faster and is has fewer effects from temperature and humidity.

Fig: Combustible gas transmitter.

The change in changing weather conditions can results dramatic changes in air quality and ambient pollution levels. The change in weather conditions are basically depends upon the factors like rates of wind dispersion i.e. their velocity as well as direction, temperature inversions, photochemical reactions, and rain.

Continuous Monitoring of Organic Pollutants

This method is used to measure very low concentrations of the volatile organic compounds in any fluid stream i.e. either liquid or gas. The fluid stream is passes through an on-line micro sorbent trap and the pollutants are concentrated or electrically heated for injection into a gas chromatograph system. The pollutants will be separated in the chromatographic column with respect to their pollution level. It is used to accurately measure the low concentrations on a continuous basis.

Advantages of Monitoring Instruments

Air quality monitoring helps us in better understanding the sources of different levels of air pollutants and effects in controlling them. It will also help in understanding the exposure of various substances in the air around us. Monitoring stations are used for continuously monitor and collect information about the presence of air pollutants. A typical monitoring station includes systematic analyzer for analyzing gaseous pollutants and their collection.

- The atmosphere is susceptible to pollution from natural sources as well as from human activities. The best way to protect air quality is to reduce the emission of pollutants by changing or improving industrial processes.
- The techniques for controlling air pollution can be either with or without pollution control equipment.
- Common types of equipment for collecting fine particulates include cyclones, scrubbers, electrostatic precipitators, and baghouse filters.
- The method for controlling of SOX is the flue-gas desulfurization techniques. In this process the fuel is burnt in such a way that the sulfur oxides are not emitted to the atmosphere.
- Flue gas treatment is used to control the oxides of nitrogen. In this process, ammonia is added to the flue gas prior to the gas passing over a catalyst.
- The other pollutants of air pollution are photochemical smog and volatile organic compounds which can be control by several chemical processes like adsorption or combustion.
- Continuous monitoring instruments basically make use of analytical methods for the determination of the concentration of air pollutants and their effect on environment.
- These instruments are available for the measurement of indoor and outdoor air pollution.
- These instruments include concentration measurement instruments, continuous emission monitoring systems, air measuring devices and meteorological instruments. The choice of monitoring instrument used for analysis will depend upon the cost of the monitoring instruments and the environment in which the monitoring equipment is required.
- In continuous emission monitoring, the system must be in continuously collects, record and report the required emissions data of air pollutants.
- It is used to accurately measure the low concentrations on a continuous basis.

References

- Carbon-Cycle, Biochemical-Cycles, Supplemental-Modules-(Environmental-Chemistry), Environmental-Chemistry: libretexts.org, Retrieved 17, May 2020
- Air-pollutants-types-sources-effects-and-control-of-air-pollutants, environment-9964: yourarticlelibrary.com, Retrieved 23, January 2020
- Acid-rain, Black, chemreview: ncsu.edu, Retrieved 11, August 2020
- Photochemical-smog, chemistry: vedantu.com, Retrieved 05, March 2020
- Planetary-energy-balance-temperature-calculate: ucar.edu, Retrieved 22, July 2020

6

Water Pollution and its Sources

The contamination of water bodies through various reasons is known as water pollution. The branch of study which focuses on the impact of manufactured chemicals and other materials either manmade or natural on aquatic animals is termed as aquatic toxicology. The topics elaborated in this chapter will help in gaining a better perspective about water pollution and toxicology as well as the different sources of water pollution.

Water Pollution

Any chemical, biological, physical change in water quality that has a harmful effect on living organisms or makes water unsuitable for desired usage.

WHO:

- 3.4 million premature deaths each year from waterborne diseases.
- 1.9 million from diarrhea.

- U.S. 1.5 million illnesses.
- 1993 Milwaukee 370,000 sick.

Infectious Agents: bacteria and viruses often from animal wastes.

Oxygen Demanding Wastes: organic waste that needs oxygen often from animal waste, paper mills and food processing.

Inorganic Chemicals: Acids and toxic chemicals often from runoff, industries and household cleaners.

Organic Chemicals: oil, gasoline, plastics, detergents often from surface runoff, industries and cleaners.

Plant Nutrients: water soluble nitrates, ammonia and phosphates often from sewage, agriculture and urban fertilizers.

Sediment: soils and silts from land erosion can disrupt photosynthesis, destroy spawning grounds, clog rivers and streams.

Heat Pollution and Radioactivity: mostly from powerplants.

How do We Measure Water Quality

Bacterial Counts: Fecal coliform counts from intestines of animals:

- None per 100 ml for drinking.
- >200 per 100 ml for swimming.

Sources: human sewage, animals, birds, raccoons, etc.

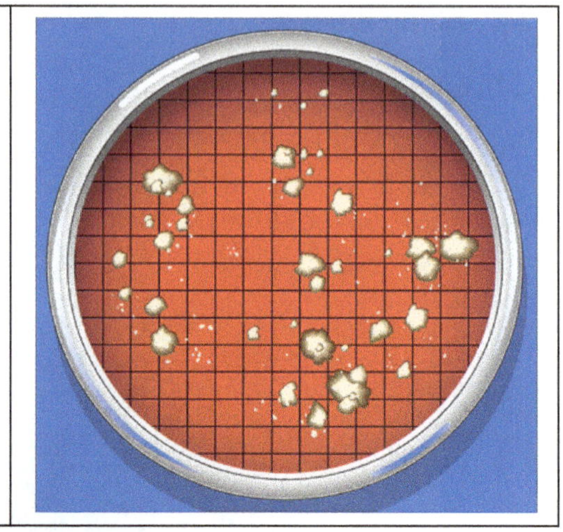

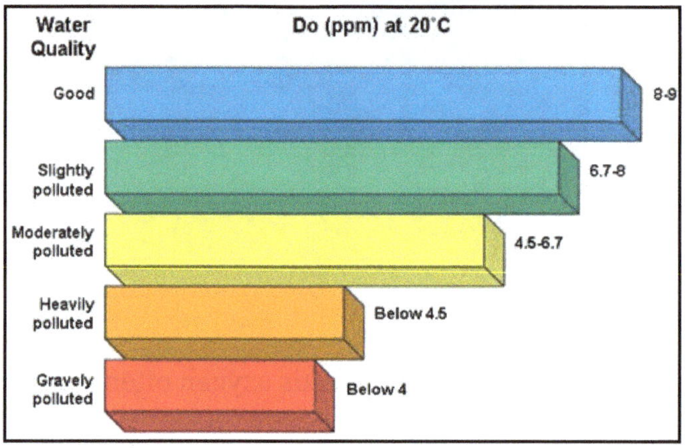

Dissolved Oxygen: BOD Biological Oxygen Demand...the amount of oxygen consumed by aquatic decomposers.

Chemical Analysis: looking for presence of inorganic or organic chemicals.

Suspended Sediment water clarity.

Indicator Species: organisms that give an idea of the health of the water body.

- Mussels, oysters and clams filter water.

Aquatic Toxicology

Aquatic toxicology is the study of the effects of manufactured chemicals and other anthropogenic and natural materials and activities on aquatic organisms at various levels of organization, from subcellular through individual organisms to communities and ecosystems. Aquatic toxicology is a multidisciplinary field which integrates toxicology, aquatic ecology and aquatic chemistry.

This field of study includes freshwater, marine water and sediment environments. Common tests include standardized acute and chronic toxicity tests lasting 24–96 hours (acute test) to 7 days or more (chronic tests). These tests measure endpoints such as survival, growth, reproduction, that are measured at each concentration in a gradient, along with a control test. Typically using selected organisms with ecologically relevant sensitivity to toxicants and a well-established literature background. These organisms can be easily acquired or cultured in lab and are easy to handle.

History

While basic research in toxicology began in multiple countries in the 1800s, it was not until around the 1930s that the use of acute toxicity testing, especially on fish, was established. Over the next two decades, the effects of chemicals and wastes on non-human species became more of a public issue and the era of the *pickle-jar bioassays* began as efforts increased to standardize toxicity testing techniques. In the United States of America, the passage of the Federal Water Pollution Control Act of 1947 marked the first comprehensive legislation for the control of water pollution and was followed by the Federal Water Pollution Control Act in 1956. In 1962, public and governmental interests were renewed, in large part due to the publication of Rachel Carson's *Silent Spring*, and three years later the Water Quality Act was passed which directed states to develop water quality standards. Public awareness, as well as scientific and governmental concern, continued to grow throughout the 1970s and by the end of the decade research had expanded to include hazard evaluation and risk analysis. In the subsequent decades, aquatic toxicology has continued to expand and internationalize so that there is now a strong application of toxicity testing for environmental protection.

Aquatic Toxicity Tests

Aquatic toxicology tests (assays): toxicity tests are used to provide qualitative and quantitative data on adverse (deleterious) effects on aquatic organisms from a toxicant. Toxicity tests can be used to assess the potential for damage to an aquatic environment and provide a database that can be used to assess the risk associated within a situation for a specific toxicant. Aquatic toxicology tests can be performed in the field or in the laboratory. Field experiments generally refer to multiple species exposure and laboratory experiments generally refer to single species exposure. A dose response relationship is most commonly used with a sigmoidal curve to quantify the toxic effects at a selected end-point or criteria for effect (i.e. death or other adverse effect to the organism). Concentration is on the x-axis and percent inhibition or response is on the y-axis.

The criteria for effects, or endpoints tested for, can include lethal and sublethal effects.

There are different types of toxicity tests that can be performed on various test species. Different species differ in their susceptibility to chemicals, most likely due to differences in accessibility, metabolic rate, excretion rate, genetic factors, dieteary factors, age, sex, health and stress level of the organism. Common standard test species are the fathead minnow (Pimephales promelas), daphnids (*Daphnia magna, D. pulex, D. pulicaria, Ceriodaphnia dubia*), midge (Chironomus tentans, C. ruparius), rainbow trout (Oncorhynchus mykiss), sheepshead minnow (Cyprinodon variegatu), mysids (Mysidopsis), oyster (Crassotreas), scud (Hyalalla Azteca), grass shrimp (Palaemonetes pugio), mussels (Mytilus). As defined by ASTM, these species are routinely selected on the basis of availability, commercial, recreational, and ecological importance, past successful use, and regulatory use.

A variety of acceptable standardized test methods have been published. Some of the more widely accepted agencies to publish methods are: the American Public Health Association, U.S. Environmental Protection Agency, American Society for Testing and Materials, International Organization for Standardization, Environment Canada, and Organization for Economic Cooperation and Development. Standardized tests offer the ability to compare results between laboratories.

There are many kinds of toxicity tests widely accepted in the scientific literature and regulatory agencies. The type of test used depends on many factors: Specific regulatory agency conducting the test, resources available, physical and chemical characteristics of the environment, type of toxicant, test species available, laboratory vs. field testing, end-point selection, and time and resources available to conduct the assays are some of the most common influencing factors on test design.

Exposure Systems

Exposure systems are four general techniques the controls and test organisms are exposed to the dealing with treated and diluted water or the test solutions.

Static- a static test exposes the organism in still water. The toxicant is added to the water in order to obtain the correct concentrations to be tested. The control and test organisms are placed in the test solutions and the water is not changed for the entirety of the test.

Recirculation- a recirculation test exposes the organism to the toxicant in a similar manner as the static test, except that the test solutions are pumped through an apparatus (i.e. filter) to maintain water quality, but not reduce the concentration of the toxicant in the water. The water is circulated through the test chamber continuously, similar to an aerated fish tank. This type of test is expensive and it is unclear whether or not the filter or aerator has an effect on the toxicant.

Renewal- a renewal test also exposes the organism to the toxicant in a similar manner as the static test because it is in still water. However, in a renewal test the test solution is renewed periodically (constant intervals) by transferring the organism to a fresh test chamber with the same concentration of toxicant.

Flow-through- a flow through test exposes the organism to the toxicant with a flow into the test chambers and then out of the test chambers. The once-through flow can either be intermittent or continuous. A stock solution of the correct concentrations of contaminant must be previously prepared. Metering pumps or diluters will control the flow and the volume of the test solution, and the proper proportions of water and contaminant will be mixed.

Types of Tests

Acute tests are short-term exposure tests (hours or days) and generally use lethality as an end-point. In acute exposures, organisms come into contact with higher doses of the toxicant in a single event or in multiple events over a short period of time and usually produce immediate effects, depending on absorption time of the toxicant. These tests are generally conducted on organisms during a specific time period of the organism's life cycle, and are considered partial life cycle tests. Acute tests are not valid if mortality in the control sample is greater than 10%. Results are reported in EC50, or concentration that will affect fifty percent of the sample size.

Chronic tests are long-term tests (weeks, months years), relative to the test organism's life span (>10% of life span), and generally use sub-lethal endpoints. In chronic exposures, organisms come into contact with low, continuous doses of a toxicant. Chronic exposures may induce effects to acute exposure, but can also result in effects that develop slowly. Chronic tests are generally considered full life cycle tests and cover an entire generation time or reproductive life cycle ("egg to egg"). Chronic tests are not considered valid if mortality in the control sample is greater than 20%. These results are generally reported in NOECs (No observed effects level) and LOECs (Lowest observed effects level).

Early life stage tests are considered as subchronic exposures that are less than a complete reproductive life cycle and include exposure during early, sensitive life stages of an organism. These exposures are also called critical life stage, embryo-larval, or egg-fry tests. Early life stage tests are not considered valid if mortality in the control sample is greater than 30%.

Short-term sublethal tests are used to evaluate the toxicity of effluents to aquatic organisms. These methods are developed by the EPA, and only focus on the most sensitive life stages. Endpoints for these test include changes in growth, reproduction and survival. NOECs, LOECs and EC50s are reported in these tests.

Bioaccumulation tests are toxicity tests that can be used for hydrophobic chemicals that may accumulated in the fatty tissue of aquatic organisms. Toxicants with low solubilities in water generally can be stored in the fatty tissue due to the high lipid content in this tissue. The storage of these toxicants within the organism may lead to cumulative toxicity. Bioaccumulation tests use bioconcentration factors (BCF) to predict concentrations of hydrophobic contaminants in organisms. The BCF is the ratio of the average concentration of test chemical accumulated in the tissue of the test organism (under steady state conditions) to the average measured concentration in the water.

Freshwater tests and saltwater tests have different standard methods, especially as set by the regulatory agencies. However, these tests generally include a control (negative and/or positive), a geometric dilution series or other appropriate logarithmic dilution series, test chambers and equal numbers of replicates, and a test organism. Exact exposure time and test duration will depend on type of test (acute vs. chronic) and organism type. Temperature, water quality parameters and light will depend on regulator requirements and organism type.

Effluent toxicity tests are tests conducted under the Clean Water Act, National Pollutant Discharge Elimination System (NPDES) permit program and are used by dischargers of contaminated effluent to monitor the quality of effluent into receiving waters. Acute Effluent Toxicity Tests are used to monitor the quality of industrial effluent monthly using acute toxicity tests. Effluent is used to perform static-acute multi concentration toxicity tests with *Ceriodaphnia dubia* and *Pimephales promelas*. The test organisms are exposed for 48 hours under static conditions with five concentrations of the effluent. Short-term Chronic Effluent Toxicity Tests are used to monitor the quality of municipal wastewater treatment plants effluent quarterly using short-term chronic toxicity tests. The goal of this test is to ensure that the wastewater is not chronically toxic. The major deviation in the short-term chronic effluent toxicity tests and the acute effluent toxicity tests is that the short-term chronic test lasts for seven days and the acute test lasts for 48 hours.

Sediment Tests

At some point most chemicals originating from both anthropogenic and natural sources accumulate in sediment. For this reason, sediment toxicity can play a major role in the adverse biological effects seen in aquatic organisms, especially those inhabiting benthic habitats. A recommended approach for sediment testing is to apply the Sediment Quality Triad (SQT) which involves simultaneously examining sediment chemistry, toxicity, and field alterations so that more complete information can be gathered. Collection, handling, and storage of sediment can have an effect on bioavailability and for this reason standard methods have been developed to suit this purpose.

Toxicological Effects

Toxicity can be broken down into two broad categories of direct and indirect toxicity. Direct toxicity results from a toxicant acting at the site of action in or on the organism. Indirect toxicity occurs with a change in the physical, chemical, or biological environment.

Lethality is most common effect used in toxicology and used as an endpoint for acute toxicity tests. While conducting chronic toxicity tests sublethal effects are endpoints that are looked at. These endpoints include behavioral, physiological, biochemical, histological changes.

There are a number of effects that occur when an organism is simultaneously exposed to two or more toxicants. These effects include additive effects, synergistic effects, potentiation effects, and antagonistic effects. An additive effect occurs when combined effect is equal to a combination or sum of the individual effects. A synergistic effect occurs when the combination of effects is much greater than the two individual effects added together. Potentiation is an effect that occurs when an individual chemical has no effect is added to a toxicant and the combination has a greater effect than just the toxicant alone. Finally, an antagonistic effect occurs when a combination of chemicals has less of an effect than the sum of their individual effects.

Streeter–phelps Equation

Example of a stream in Arkhangelsk Oblast, Russia.

Example of a river, Tigris River near Hasankeyf, in Turkey.

The Streeter–Phelps equation is used in the study of water pollution as a water quality modelling tool. The model describes how dissolved oxygen (DO) decreases in a river or stream along a certain distance by degradation of biochemical oxygen demand (BOD). The equation was derived by H. W. Streeter, a sanitary engineer, and Earle B. Phelps, a consultant for the U.S. Public Health Service, in 1925, based on field data from the Ohio River. The equation is also known as the DO sag equation.

The Streeter–Phelps equation determines the relation between the dissolved oxygen concentration and the biological oxygen demand over time and is a solution to the linear first order differential equation:

$$\frac{\partial D}{\partial t} = k_1 L_t - k_2 D$$

This differential equation states that the total change in oxygen deficit (D) is equal to the difference between the two rates of deoxygenation and reaeration at any time.

The Streeter–Phelps equation, assuming a plug-flow stream at steady state is then:

$$D = \frac{k_1 L_a}{k_2 - k_1}(e^{-k_1 t} - e^{-k_2 t}) + D_a e^{-k_2 t}$$

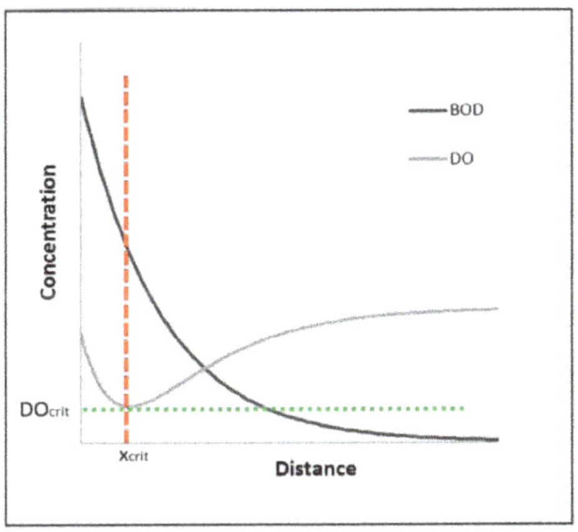

Streeter-Phelps DO sag curve and BOD development.

where

- D is the saturation deficit, which can be derived from the dissolved oxygen concentration at saturation minus the actual dissolved oxygen concentration ($D = DO_{sat} - DO$). D has the dimensions $\frac{g}{m^3}$.
- k_1 is the deoxygenation rate, usually in d^{-1}.
- k_2 is the reaeration rate, usually in d^{-1}.
- L_a is the initial oxygen demand of organic matter in the water, also called the ultimate BOD (BOD at time t=infinity). The unit of L_a is $\frac{g}{m^3}$.
- L_t is the oxygen demand remaining at time t, $L_t = L_a e^{-k_1 t}$.
- D_a is the initial oxygen deficit $[\frac{g}{m^3}]$.
- t is the elapsed time, usually $[d]$.

k_1 lies typically within the range 0.05-0.5 d^{-1} and k_2 lies typically within the range 0.4-1.5 d^{-1}. The Streeter–Phelps equation is also known as the DO sag equation. This is due to the shape of the graph of the DO over time.

Critical Oxygen Deficit

On the DO sag curve a minimum concentration occurs at some point, along a stream. If the Streeter–Phelps equation is differentiated with respect to time, and set equal to zero, the time at which the minimum DO occurs is expressed by:

$$t_{crit} = \frac{1}{k_2 - k_1} \ln\left[\frac{k_2}{k_1}\left(1 - \frac{D_a(k_2 - k_1)}{L_a k_1}\right)\right]$$

To find the value of the critical oxygen deficit, D_{crit}, the Streeter–Phelps equation is combined with the equation above, for the critical time, t_{crit}. Then the minimum dissolved oxygen concentration is:

$$DO_{crit} = DO_{sat} - D_{crit}$$

It should be noted that, mathematically it is possible to get a negative value of DO_{crit}, even though it is not possible to have a negative amount of DO in reality.

The distance traveled in a river from a given point source pollution or waste discharge downstream to the DO_{crit} (which is the minimum DO) is found by:

$$x_{crit} = vt_{crit}$$

where v is the flow velocity of the stream. This formula is a good approximation as long as the flow can be regarded as a plug flow (turbulent).

Estimation of Reaeration Rate

Several estimations of the reaeration rate exist, which generally follow the equation:

$$k_2 = Kv^a H^{-b}$$

where

- K is a constant.
- v is the flow velocity [m/s].
- H is the depth [m].
- a is a constant.
- b is a constant.

The constants depend on the system to which the equation is applied, i.e. the flow velocity and the size of the stream or river. Different values are available in the literature.

The software "International Hydrological Programme" applies the following equation derived on the basis of values used in published literature:

$$k_2 = 2.148v^{0.878}H^{-1.48}$$

where

- k_2 [d^{-1}].
- v is the average flow velocity [m/s].
- H is the average depth of flow in the river [m].

Temperature Correction

Both the deoxygenation rate, k_1 and reaeration rate, k_2 can be temperature corrected, following the general formula.

$$k = k_{20}\theta^{(T-20)}$$

where

- k_{20} is the rate at 20 degrees Celsius.
- θ is a constant, which differs for the two rates.
- is the actual temperature in the stream in degC.

Normally θ has the value 1.048 for k_1 and 1.024 for k_2. An increasing temperature has the most impact on the deoxygenation rate, and results in an increased critical deficit (D_{crit}), and x_{crit} decreases. Furthermore, a decreased DO_{sat} concentration occurs with increasing temperature, which leads to a decrease in the DO concentration.

Mixing of Rivers

When two streams or rivers merge or water is discharged to a stream it is possible to determine the BOD and DO after mixing assuming steady state conditions and instantaneous mixing. The two streams are considered as dilutions of each other thus the initial BOD and DO will be:

$$L_a = \frac{L_s Q_s + L_b Q_b}{Q_s + Q_b}$$

and

$$DO_0 = \frac{DO_s Q_s + DO_b Q_b}{Q_s + Q_b}$$

where

- L_a is the initial concentration of BOD in the river downstream of the mixing, also called BOD(0). The unit of L_a is $\frac{g}{m^3}$.
- L_b is the background BOD of the concentration in the river [$\frac{g}{m^3}$].

- L_s is the BOD of the content of the merging river $[\frac{g}{m^3}]$.
- DO_0 is the initial concentration of the dissolved oxygen in the river downstream of the conjoining point $[\frac{g}{m^3}]$.
- DO_b is the background concentration of the dissolved oxygen content in the river $[\frac{g}{m^3}]$.
- DO_s is the background concentration of the dissolved oxygen content in the merging river $[\frac{g}{m^3}]$.
- Q_b is the flow in the river upstream from the mixing point $[\frac{m^3}{s}]$.
- Q_s is the flow in the merging river upstream from the mixing point $[\frac{m^3}{s}]$.

Numerical Approach

Nowadays it is possible to solve the classical Streeter–Phelps equation numerically by use of computers. The differential equations are solved by integration.

History

In 1925, a study on the phenomena of oxidation and reaeration in the Ohio River in the US was published by the sanitary engineer, Harold Warner Streeter and the consultant, Earle Bernard Phelps (1876–1953). The study was based on data obtained from May 1914 to April 1915 by the United States Public Health Service under supervision of Surg. W.H. Frost.

More complex versions of the Streeter-Phelps model were introduced during the 1960s, where computers made it possible to include further contributions to the oxygen development in streams. At the head of this development were O'Connor (1960) and Thomann (1963). O'Connor added the contributions from photosynthesis, respiration and sediment oxygen demand (SOD). Thomann expanded the Streeter-Phelps model to allow for multi segment systems.

Applications and Limitations

The simple Streeter-Phelps model is based on the assumptions that a single BOD input is distributed evenly at the cross section of a stream or river and that it moves as plug flow with no mixing in the river. Furthermore only one DO sink (carbonaceous BOD) and one DO source (reaeration) is considered in the classical Streeter-Phelps model. These simplifications will give rise to errors in the model. For example the model does not include BOD removal by sedimentation, that suspended BOD is converted to a dissolved state, that sediment has an oxygen demand and that photosynthesis and respiration will impact the oxygen balance.

Expanded Model

In addition to the oxidation of organic matter and the reaeration process, there are many other processes in a stream which affect the DO. In order to make a more accurate model it is possible to include these factors using an expanded model.

The expanded model is a modification of the traditional model and includes internal sources (re-aeration and photosynthesis) and sinks (BOD, background BOD, SOD and respiration) of DO. It is not always necessary to include all of these parameters. Instead relevant sources and sinks can be summed to yield the overall solution for the particular model. Parameters in the expanded model can be either measured in the field or estimated theoretically.

Background BOD

Background BOD or benthic oxygen demand is the diffuse source of BOD represented by the decay of organic matter that has already settled on the bottom. This will give rise to a constant diffuse input thus the change in BOD over time will be:

$$\frac{dL}{dt} = -k_1 L + L_b$$

where

- k_1 is the rate for oxygen consumption by BOD, usually in d^{-1}.
- L is the BOD from organic matter in the water [—].
- L_b is the background BOD input $[\frac{\text{g}}{\text{m}^3}]$.

Sedimentation of BOD

Sedimented BOD does not directly consume oxygen and this should therefore be taken into account. This is done by introducing a rate of BOD removal combined with a rate of oxygen consumption by BOD. Giving a total rate for oxygen removal by BOD:

$$k_r = k_1 + k_3$$

where

- k_1 is the rate of oxygen consumption by BOD, usually in d^{-1}.
- k_3 is the rate of settling of BOD, usually in d^{-1}.

The change in BOD over time is described as:

$$\frac{dL}{dt} = -k_r L$$

where L is the BOD from organic matter in the water $[\frac{\text{g}}{\text{m}^3}]$.

k_r is typically in the range of 0.5-5 d^{-1}.

Sediment Oxygen Demand

Oxygen can be consumed by organisms in the sediment. This process is referred to as sediment oxygen demand (SOD). Measurement of SOD can be undertaken by measuring the change of oxygen in a box on the sediment.(benthic respirometer).

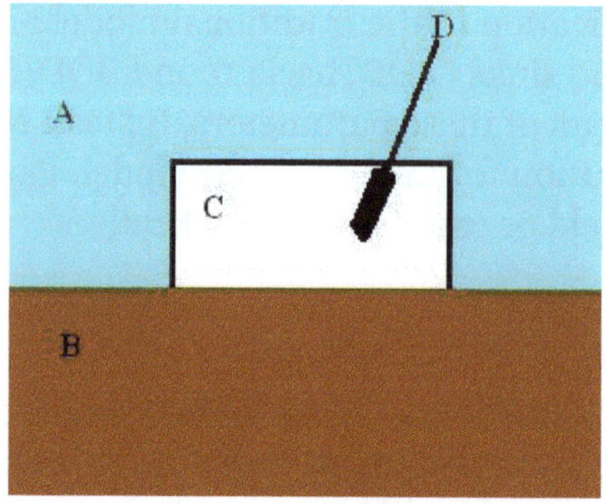

Measurement of SOD using an oxygen electrode. A: Water, B: Sediment, C: Box, D: Oxygen electrode.

The change in oxygen deficit due to consumption by sediment is described as:

$$\frac{dD}{dt} = -k_2 D + \frac{S}{H}$$

where

- H is the depth of the river [m].
- S is the SOD $[\frac{g}{m^2 d}]$.
- D is the saturation deficit $[\frac{g}{m^3}]$.
- k_2 is the reaeration rate $[d^{-1}]$.

The range of the SOD is typically in the range of 0.1 – 1 $\frac{g}{m^2 d}$ for a natural river with low pollution and 5 – 10 $\frac{g}{m^2 d}$ for a river with moderate to heavy pollution.

Nitrification

Ammonium is oxidized to nitrate under aerobic conditions:

$$NH_4^+ + 2O_2 \rightarrow NO_3^- + H_2O + 2H^+$$

Ammonium oxidation can be treated as part of BOD, so that BOD = CBOD + NBOD, where CBOD is the carbonaceous biochemical oxygen demand and NBOD is nitrogenous BOD. Usually CBOD is much higher than the ammonium concentration and thus NBOD often does not need to be considered. The change in oxygen deficit due to oxidation of ammonium is described as:

$$\frac{dD}{dt} = k_N N - k_2 D$$

where

- D is the saturation deficit.
- k_N is the nitrification rate $[d^{-1}]$.
- N is ammonium-nitrogen concentration.

The range of k_N is typically 0.05-0.5 d^{-1}.

Photosynthesis and Respiration

Photosynthesis and respiration are performed by algae and by macrophytes. Respiration is also performed by bacteria and animals. Assuming steady state (net daily average) the change in deficit will be:

$$\frac{dD}{dt} = -k_2 D + (R+P)_{avg}$$

where

- R is the respiration $[\frac{mg}{Ld}]$.
- P is the photosynthesis $[\frac{mg}{Ld}]$.

Note that BOD only includes respiration of microorganisms e.g. algae and bacteria and not by macrophytes and animals.

Due to the variation of light over time, the variation of the photosynthetic oxygen can be described by a periodical function over time, where time is after sunrise and before sunset:

$$P(t) = P_{max} sin\left(\frac{\pi}{f} + (t - t_s)\right)$$

where

- $P(t)a$ is the photosynthesis at a given time $[\frac{mg}{Ld}]$.
- P_{max} is the daily maximum of the photosynthesis $[\frac{mg}{Ld}]$.
- f is the fraction of day with sunlight, usually $\frac{1}{2}$ day.
- t_s is the time at which sun rises $[d]$.

The range of the daily average value of primary production $(P-R)$ is typically 0.5-10 $\frac{mg}{Ld}$.

This contribution was made originally as part of the Water Pollution course at the Technical University of Denmark in 2011.

Sources of Water Pollution

Types, Effects and Sources of Water Pollution:

- Point sources.
- Nonpoint sources.
- Water quality.

Nonpoint Source Pollution

Nonpoint source (NPS) pollution is a term used to describe pollution resulting from many diffuse sources, in direct contrast to point source pollution which results from a single source. Nonpoint source pollution generally results from land runoff, precipitation, atmospheric deposition, drainage, seepage, or hydrological modification (rainfall or snowmelt) where tracing the pollution back to a single source is difficult.

Muddy river.

Non-point source water pollution affects a water body from sources such as polluted runoff from agricultural areas draining into a river, or wind-borne debris blowing out to sea. Non-point source air pollution affects air quality from sources such as smokestacks or car tailpipes. Although these pollutants have originated from a point source, the long-range transport ability and multiple sources of the pollutant make it a non-point source of pollution. Non-point source pollution can be contrasted with point source pollution, where discharges occur to a body of water or into the atmosphere at a single location.

NPS may derive from many different sources with no specific solution may change to rectify the problem, making it difficult to regulate. Non point source water pollution is difficult to control because it comes from the everyday activities of many different people, such as fertilizing a lawn, using a pesticide, or constructing a road or building.

It is the leading cause of water pollution in the United States today, with polluted runoff from agriculture the primary cause.

Other significant sources of runoff include hydrological and habitat modification, and silviculture (forestry).

Contaminated stormwater washed off parking lots, roads and highways, and lawns (often containing fertilizers and pesticides) is called urban runoff. This runoff is often classified as a type of NPS pollution. Some people may also consider it a point source because many times it is channeled into municipal storm drain systems and discharged through pipes to nearby surface waters. However, not all urban runoff flows through storm drain systems before entering water bodies. Some may flow directly into water bodies, especially in developing and suburban areas. Also, unlike other types of point sources, such as industrial discharges, sewage treatment plants and other operations, pollution in urban runoff cannot be attributed to one activity or even group of activities. Therefore, because it is not caused by an easily identified and regulated activity, urban runoff pollution sources are also often treated as true non-point sources as municipalities work to abate them.

Principal Types

Runoff of soil and fertilizer during a rain storm.

Sediment

Sediment (loose soil) includes silt (fine particles) and suspended solids (larger particles). Sediment may enter surface waters from eroding stream banks, and from surface runoff due to improper plant cover on urban and rural land. Sediment creates turbidity (cloudiness) in water bodies, reducing the amount of light reaching lower depths, which can inhibit growth of submerged aquatic plants and consequently affect species which are dependent on them, such as fish and shellfish. High turbidity levels also inhibit drinking water purification systems.

Sediment can also be discharged from multiple different sources. Sources include construction sites (although these are point sources, which can be managed with erosion controls and sediment controls), agricultural fields, stream banks, and highly disturbed areas.

Nutrients

Nutrients mainly refers to inorganic matter from runoff, landfills, livestock operations and crop lands. The two primary nutrients of concern are phosphorus and nitrogen.

Phosphorus is a nutrient that occurs in many forms that are bioavailable. It is notoriously over-abundant in human sewage sludge. It is a main ingredient in many fertilizers used for agriculture as well as on residential and commercial properties, and may become a limiting nutrient in freshwater systems and some estuaries. Phosphorus is most often transported to water bodies via soil erosion because many forms of phosphorus tend to be adsorbed on to soil particles. Excess amounts of phosphorus in aquatic systems (particularly freshwater lakes, reservoirs, and ponds) leads to proliferation of microscopic algae called phytoplankton. The increase of organic matter supply due to the excessive growth of the phytoplankton is called eutrophication. A common symptom of eutrophication is algae blooms that can produce unsightly surface scums, shade out beneficial types of plants, and poison the water due to toxins produced by the algae. These toxins are a particular problem in systems used for drinking water because some toxins can cause human illness and removal of the toxins is difficult and expensive. Bacterial decomposition of algal blooms consumes dissolved oxygen in the water, generating hypoxia with detrimental consequences for fish and aquatic invertebrates.

Nitrogen is the other key ingredient in fertilizers, and it generally becomes a pollutant in saltwater or brackish estuarine systems where nitrogen is a limiting nutrient. Similar to phosphorus in fresh-waters, excess amounts of bioavailable nitrogen in marine systems lead to eutrophication and algae blooms. Hypoxia is an increasingly common result of eutrophication in marine systems and can impact large areas of estuaries, bays, and near shore coastal waters. Each summer, hypoxic conditions form in bottom waters where the Mississippi River enters the Gulf of Mexico. During recent summers, the aereal extent of this "dead zone" is comparable to the area of New Jersey and has major detrimental consequences for fisheries in the region.

Nitrogen is most often transported by water as nitrate (NO_3). The nitrogen is usually added to a watershed as organic-N or ammonia (NH_3), so nitrogen stays attached to the soil until oxidation converts it into nitrate. Since the nitrate is generally already incorporated into the soil, the water traveling through the soil (i.e., interflow and tile drainage) is the most likely to transport it, rather than surface runoff.

Toxic Contaminants and Chemicals

Compounds including heavy metals like lead, mercury, zinc, and cadmium, organics like polychlorinated biphenyls (PCBs) and polycyclic aromatic hydrocarbons (PAHs), fire retardants, and other substances are resistant to breakdown. These contaminants can come from a variety of sources including human sewage sludge, mining operations, vehicle emissions, fossil fuel combustion, urban runoff, industrial operations and landfills.

Toxic chemicals mainly include organic compounds and inorganic compounds. These compounds include pesticides like DDT, acids, and salts that have severe effects to the ecosystem and water-bodies. These compounds can threaten the health of both humans and aquatic species while being resistant to environmental breakdown, thus allowing them to persist in the environment. These toxic chemicals could come from croplands, nurseries, orchards, building sites, gardens, lawns and landfills.

Acids and salts mainly are inorganic pollutants from irrigated lands, mining operations, urban runoff, industrial sites and landfills.

Pathogens

Pathogens are bacteria and viruses that can be found in water and cause diseases in humans. Typically, pathogens cause disease when they are present in public drinking water supplies. Pathogens found in contaminated runoff may include:

- *Cryptosporidium parvum*
- *Giardia lamblia*
- *Salmonella*
- *Novovirus* and other viruses
- Parasitic worms (helminths)

Coliform bacteria and fecal matter may also be detected in runoff. These bacteria are a commonly used indicator of water pollution, but not an actual cause of disease.

Pathogens may contaminate runoff due to poorly managed livestock operations, faulty septic systems, improper handling of pet waste, the over application of human sewage sludge, contaminated storm sewers, and sanitary sewer overflows.

Principal Sources

Urban and Suburban Areas

Urban and suburban areas are a main sources of nonpoint source pollution due to the amount of runoff that is produced due to the large amount of paved surfaces. Paved surfaces, such as asphalt and concrete are impervious to water penetrating them. Any water that is on contact with these surfaces will run off and be absorbed by the surrounding environment. These surfaces make it easier for stormwater to carry pollutants into the surrounding soil.

Construction sites tend to have disturbed soil that is easily eroded by precipitation like rain, snow, and hail. Additionally, discarded debris on the site can be carried away by runoff waters and enter the aquatic environment.

Typically, in suburban areas, chemicals are used for lawn care. These chemicals can end up in runoff and enter the surrounding environment via storm drains in the city. Since the water in storm drains is not treated before flowing into surrounding water bodies, the chemicals enter the water directly.

Agricultural Operations

Agricultural operations account for a large percentage of all nonpoint source pollution in the United States. When large tracts of land are plowed to grow crops, it exposes and loosens soil that was once buried. This makes the exposed soil more vulnerable to erosion during rainstorms. It also can increase the amount of fertilizer and pesticides carried into nearby bodies of water.

Atmospheric Inputs

Atmospheric inputs of pollution into the air can come from multiple sources. Typically, industrial facilities, like factories, emit air pollution via a smokestack. Although this is a point source, due to the distributional nature, long-range transport, and multiple sources of the pollution, it is considered a nonpoint source. Additionally, atmospheric pollution can become water pollution, by being washed out of the atmosphere in the form of rain or snow.

Highway Runoff

Highway runoff accounts for a small but widespread percentage of all nonpoint source pollution. Harned (1988) estimated that runoff loads were composed of atmospheric fallout (9%), vehicle deposition (25%) and highway maintenance materials (67%) he also estimated that about 9 percent of these loads were reentrained in the atmosphere.

Forestry and Mining Operations

Forestry and mining operations can have significant inputs to non-point source pollution.

Forestry

Forestry operations reduce the number of trees in a given area, thus reducing the oxygen levels in that area as well. This action, coupled with the heavy machinery rolling over the soil increases the risk of erosion.

Mining

Active mining operations are considered point sources, however runoff from abandoned mining operations contribute to nonpoint source pollution. In strip mining operations, the top of the mountain is removed to expose the desired ore. If this area is not properly reclaimed once the mining has finished, soil erosion can occur. Additionally, there can be chemical reactions with the air and newly exposed rock to create acidic runoff. Water that seeps out of abandoned subsurface mines can also be highly acidic. This can seep into the nearest body of water and change the pH in the aquatic environment.

Marinas and Boating Activities

Chemicals used for boat maintenance, like paint, solvents, and oils find their way into water through runoff. Additionally, spilling fuels or leaking fuels directly into the water from boats contribute to nonpoint source pollution. Nutrient and bacteria levels are increased by poorly maintained sanitary waste receptacles on the boat and pump-out stations.

Control

Contour buffer strips used to retain soil and reduce erosion.

Urban and Suburban Areas

To control non-point source pollution, many different approaches can be undertaken in both urban and suburban areas. Buffer strips provide a barrier of grass in between impervious paving material like parking lots and roads, and the closest body of water. This allows the soil to absorb any pollution before it enters the local aquatic system. Retention ponds can be built in drainage areas to create an aquatic buffer between runoff pollution and the aquatic environment. Runoff and storm water drain into the retention pond allowing for the contaminants to settle out and become trapped in the pond.

The use of porous pavement allows for rain and storm water to drain into the ground beneath the pavement, reducing the amount of runoff that drains directly into the water body. Restoration methods such as constructing wetlands are also used to slow runoff as well as absorb contamination.

Construction sites typically implement simple measures to reduce pollution and runoff. Firstly, sediment or silt fences are erected around construction sites to reduce the amount of sediment and large material draining into the nearby water body. Secondly, laying grass or straw along the border of construction sites also work to reduce nonpoint source pollution.

In areas served by single-home septic systems, local government regulations can force septic system maintenance to ensure compliance with water quality standards. In Washington (state), a novel approach was developed through a creation of a "shellfish protection district" when either a commercial or recreational shellfish bed is downgraded because of ongoing nonpoint source pollution. The shellfish protection district is a geographic area designated by a county to protect water quality and tideland resources, and provides a mechanism to generate local funds for water quality services to control nonpoint sources of pollution. At least two shellfish protection districts in south Puget Sound have instituted septic system operation and maintenance requirements with program fees tied directly to property taxes.

Agricultural Operations

To control sediment and runoff, farmers may utilize erosion controls to reduce runoff flows and retain soil on their fields. Common techniques include contour plowing, crop mulching, crop rotation, planting perennial crops and installing riparian buffers. *Conservation tillage* is a concept used to reduce runoff while planting a new crop. The farmer leaves some crop reside from the previous planting in the ground to help prevent runoff during the planting process.

Nutrients are typically applied to farmland as commercial fertilizer; animal manure; or spraying of municipal or industrial wastewater (effluent) or sludge. Nutrients may also enter runoff from crop residues, irrigation water, wildlife, and atmospheric deposition. Farmers can develop and implement nutrient management plans to reduce excess application of nutrients.

To minimize pesticide impacts, farmers may use Integrated Pest Management (IPM) techniques (which can include biological pest control) to maintain control over pests, reduce reliance on chemical pesticides, and protect water quality.

Forestry Operations

With a well-planned placement of both logging trails, also called skid trails, can reduce the amount of sediment generated. By planning the trails location as far away from the logging activity as possible as well as contouring the trails with the land, it can reduce the amount of loose sediment in the runoff. Additionally, by replanting trees on the land after logging, it provides a structure for the soil to regain stability as well as replaces the logged environment.

Marinas

Installing shut off valves on fuel pumps at a marina dock can help reduce the amount of spillover into the water. Additionally, pump-out stations that are easily accessible to boaters in a marina can provide a clean place in which to dispose of sanitary waste without dumping it directly into the water. Finally, something as simple as having trash containers around a marina can prevent larger objects entering the water.

Groundwater Pollution

Groundwater pollution example in Lusaka, Zambia where the pit latrine in the background is polluting the shallow well in the foreground with pathogens and nitrate.

Groundwater pollution (also called groundwater contamination) occurs when pollutants are released to the ground and make their way down into groundwater. It can also occur naturally due to the presence of a minor and unwanted constituent, contaminant or impurity in the groundwater, in which case it is more likely referred to as contamination rather than pollution.

The pollutant creates a contaminant plume within an aquifer. Movement of water and dispersion within the aquifer spreads the pollutant over a wider area. Its advancing boundary, often called a plume edge, can intersect with groundwater wells or daylight into surface water such as seeps and spring, making the water supplies unsafe for humans and wildlife. The movement of the plume, called a plume front, may be analyzed through a hydrological transport model or groundwater model. Analysis of groundwater pollution may focus on soil characteristics and site geology, hydrogeology, hydrology, and the nature of the contaminants.

Pollution can occur from on-site sanitation systems, landfills, effluent from wastewater treatment plants, leaking sewers, petrol filling stations or from over application of fertilizers in agriculture. Pollution (or contamination) can also occur from naturally occurring contaminants, such as arsenic or fluoride. Using polluted groundwater causes hazards to public health through poisoning or the spread of disease.

Different mechanisms have influence on the transport of pollutants, e.g. diffusion, adsorption, precipitation, decay, in the groundwater. The interaction of groundwater contamination with surface waters is analyzed by use of hydrology transport models.

Pollutant Types

Contaminants found in groundwater cover a broad range of physical, inorganic chemical, organic chemical, bacteriological, and radioactive parameters. Principally, many of the same pollutants that play a role in surface water pollution may also be found in polluted groundwater, although their respective importance may differ.

Arsenic and Fluoride

Arsenic and fluoride have been recognized by the World Health Organization (WHO) as the most serious inorganic contaminants in drinking-water on a worldwide basis.

The metalloid arsenic can occur naturally in groundwater, as seen most frequently in Asia, including in China, India and Bangladesh. In the Ganges Plain of northern India and Bangladesh severe contamination of groundwater by naturally occurring arsenic affects 25% of water wells in the shallower of two regional aquifers.

Arsenic in groundwater can also be present where there are mining operations or mine waste dumps that will leach arsenic.

Natural fluoride in groundwater is of growing concern as deeper groundwater is being used, "with more than 200 million people at risk of drinking water with elevated concentrations." Fluoride can especially be released from acidic volcanic rocks and dispersed volcanic ash when water hardness is low. High levels of fluoride in groundwater is a serious problem in the Argentinean Pampas, Chile, Mexico, India, Pakistan, the East African Rift, and some volcanic islands (Tenerife).

In areas that have naturally occurring high levels of fluoride in groundwater which is used for drinking water, both dental and skeletal fluorosis can be prevalent and severe.

Pathogens

Waterborne diseases can be spread via a groundwater well which is contaminated with fecal pathogens from pit latrines.

Pathogens contained in feces can lead to groundwater pollution when they are given the opportunity to reach the groundwater, making it unsafe for drinking. Of the four pathogen types that are present in feces (bacteria, viruses, protozoa and helminths or helminth eggs), the first three can be commonly found in polluted groundwater, whereas the relatively large helminth eggs are usually filtered out by the soil matrix.

Groundwater that is contaminated with pathogens can lead to fatal fecal-oral transmission of diseases (e.g. cholera, diarrhoea).

Nitrate

Nitrate is the most common chemical contaminant in the world's groundwater and aquifers. In some low-income countries nitrate levels in groundwater are extremely high, causing significant health problems. It is also stable (it does not degrade) under high oxygen conditions.

Nitrate levels above 10 mg/L (10 ppm) in groundwater can cause "blue baby syndrome" (acquired methemoglobinemia). Drinking water quality standards in the European Union stipulate less than 50 mg/L for nitrate in drinking water.

However, the linkages between nitrates in drinking water and blue baby syndrome have been disputed in other studies. The syndrome outbreaks might be due to other factors than elevated nitrate concentrations in drinking water.

Elevated nitrate levels in groundwater can be caused by on-site sanitation, sewage sludge disposal and agricultural activities. It can therefore have an urban or agricultural origin.

Organic Compounds

Volatile organic compounds (VOCs) are a dangerous contaminant of groundwater. They are generally introduced to the environment through careless industrial practices. Many of these compounds were not known to be harmful until the late 1960s and it was some time before regular testing of groundwater identified these substances in drinking water sources.

Primary VOC pollutants found in groundwater include aromatic hydrocarbons such as BTEX compounds (benzene, toluene, ethylbenzene and xylenes), and chlorinated solvents including tetrachloroethylene (PCE), trichloroethylene (TCE), and vinyl chloride (VC). BTEX are important components of gasoline. PCE and TCE are industrial solvents historically used in dry cleaning processes and as a metal degreaser, respectively.

Other organic pollutants present in groundwater and derived from industrial operations are the polycyclic aromatic hydrocarbons (PAHs). Due to its molecular weight, Naphthalene is the most soluble and mobile PAH found in groundwater, whereas benzo(a)pyrene is the most toxic one. PAHs are generally produced as byproducts by incomplete combustion of organic matter.

Organic pollutants can also be found in groundwater as insecticides and herbicides. As many other synthetic organic compounds, most pesticides have very complex molecular structures. This complexity determines the water solubility, adsorption capacity, and mobility of pesticides in the groundwater system. Thus, some types of pesticides are more mobile than others so they can more easily reach a drinking-water source.

Metals

Several trace metals can occurs naturally in some rocks or enter in the environmental from natural processes such as weathering. However, industrial activities such as mining and metallurgy, solid waste disposal, painting and enamel works can lead to severe groundwater pollution with elevated concentrations of toxic metals including lead, cadmium and chromium.

The migration of metals (and metalloids) in groundwater will be affected by several factors, in particular by chemical reactions which determine the partitioning of contaminants among different phases and species. Thus, the mobility of metals primarily depends on the pH and redox state of groundwater.

Others

Other organic pollutants include a range of organohalides and other chemical compounds, petroleum hydrocarbons, various chemical compounds found in personal hygiene and cosmetic products, drug pollution involving pharmaceutical drugs and their metabolites. Inorganic pollutants might include other nutrients such as ammonia and phosphate, and radionuclides such as uranium (U) or radon (Rn) naturally present in some geological formations. Saltwater intrusion is also an example of natural contamination, but is very often intensified by human activities.

Groundwater pollution is a worldwide issue. A study of the groundwater quality of the principal aquifers of the United States conducted between 1991 and 2004, showed that 23% of domestic wells had contaminants at levels greater than human-health benchmarks. Another study suggested that the major groundwater pollution problems in Africa, considering the order of importance are: (1) nitrate pollution, (2) pathogenic agents, (3) organic pollution, (4) salinization, and (5) acid mine drainage.

Causes

Naturally-occurring (Geogenic)

"Geogenic" refers to naturally occurring as a result from geological processes.

The natural arsenic pollution occurs because aquifer sediments contain organic matter that generates anaerobic conditions in the aquifer. These conditions result in the microbial dissolution of iron oxides in the sediment and, thus, the release of the arsenic, normally strongly bound to iron oxides, into the water. As a consequence, arsenic-rich groundwater is often iron-rich, although secondary processes often obscure the association of dissolved arsenic and dissolved iron. Arsenic is found in groundwater most commonly as the reduced species arsenite and the oxidized species arsenate, being the acute toxicity of arsenite somewhat greater than that of arsenate. Investigations by WHO indicated that 20% of 25,000 boreholes tested in Bangladesh had arsenic concentrations exceeding 50 μg/l.

The occurrence of fluoride is close related to the abundance and solubility of fluoride-containing minerals such as fluorite (CaF2). Considerably high concentrations of fluoride in groundwater are typically caused by a lack of calcium in the aquifer. Health problems associated with dental fluorosis may occur when fluoride concentrations in groundwater exceed 1.5 mg/l, which is the WHO guideline value since 1984.

The Swiss Federal Institute of Aquatic Science and Technology (EAWAG) has recently developed the interactive Groundwater Assessment Platform (GAP), where the geogenic risk of contamination in a given area can be estimated using geological, topographical and other environmental data without having to test samples from every single groundwater resource. This tool also allows the user to produce probability risk mapping for both arsenic and fluoride.

High concentrations of parameters like salinity, iron, manganese, uranium, radon and chromium, in groundwater, may also be of geogenic origin. This contaminants can be important locally but they are not as widespread as arsenic and fluoride.

On-site Sanitation Systems

A traditional housing compound near Herat, Afghanistan, where a shallow water supply well (foreground) is in close proximity to the pit latrine (behind the white greenhouse) leading to contamination of the groundwater.

Groundwater pollution with pathogens and nitrate can also occur from the liquids infiltrating into the ground from on-site sanitation systems such as pit latrines and septic tanks, depending on the population density and the hydrogeological conditions.

Factors controlling the fate and transport of pathogens are quite complex and the interaction among them is not well understood. If the local hydrogeological conditions (which can vary within a space of a few square kilometres) are ignored, simple on-site sanitation infrastructures such as pit latrines can cause significant public health risks via contaminated groundwater.

Liquids leach from the pit and pass the unsaturated soil zone (which is not completely filled with water). Subsequently, these liquids from the pit enter the groundwater where they may lead to groundwater pollution. This is a problem if a nearby water well is used to supply groundwater for drinking water purposes. During the passage in the soil, pathogens can die off or be adsorbed significantly, mostly depending on the travel time between the pit and the well. Most, but not all pathogens die within 50 days of travel through the subsurface.

The degree of pathogen removal strongly varies with soil type, aquifer type, distance and other environmental factors. For example, the unsaturated zone becomes "washed" during extended periods of heavy rain, providing hydraulic pathway for the quick pass of pathogens. It is difficult to estimate the safe distance between a pit latrine or a septic tank and a water source. In any case, such recommendations about the safe distance are mostly ignored by those building pit latrines. In addition, household plots are of a limited size and therefore pit latrines are often built much closer to groundwater wells than what can be regarded as safe. This results in groundwater pollution and household members falling sick when using this groundwater as a source of drinking water.

Sewage (Treated and Untreated)

Groundwater pollution can be caused by untreated waste discharge leading to diseases like skin lesions, bloody diarrhea and dermatitis. This is more common in locations having limited wastewater treatment infrastructure, or where there are systematic failures of the on-site sewage disposal system. Along with pathogens and nutrients, untreated sewage can also have an important load of heavy metals that may seep into the groundwater system.

The treated effluent from sewage treatment plants may also reach the aquifer if the effluent is infiltrated or discharged to local surface water bodies. Therefore, those substances that are not removed in conventional sewage treatment plants may reach the groundwater as well. For example, detected concentrations of pharmaceutical residues in groundwater were in the order of 50 ng/L in several locations in Germany. This is because in conventional sewage treatment plants, micro-pollutants such as hormones, pharmaceutical residues and other micro-pollutants contained in urine and feces are only partially removed and the remainder is discharged into surface water, from where it may also reach the groundwater.

Groundwater pollution can also occur from leaking sewers which has been observed for example in Germany. This can also lead to potential cross-contamination of drinking-water supplies.

Spreading wastewater or sewage sludge in agriculture may also be included as sources of faecal contamination in groundwater.

Fertilizers and Pesticides

Nitrate can also enter the groundwater via excessive use of fertilizers, including manure spreading. This is because only a fraction of the nitrogen-based fertilizers is converted to produce and other plant matter. The remainder accumulates in the soil or lost as run-off. High application rates of nitrogen-containing fertilizers combined with the high water-solubility of nitrate leads to increased runoff into surface water as well as leaching into groundwater, thereby causing groundwater pollution. The excessive use of nitrogen-containing fertilizers (be they synthetic or natural) is particularly damaging, as much of the nitrogen that is not taken up by plants is transformed into nitrate which is easily leached.

The nutrients, especially nitrates, in fertilizers can cause problems for natural habitats and for human health if they are washed off soil into watercourses or leached through soil into groundwater. The heavy use of nitrogenous fertilizers in cropping systems is the largest contributor to anthropogenic nitrogen in groundwater worldwide.

Feedlots/animal corrals can also lead to the potential leaching of nitrogen and metals to groundwater. Over application of animal manure may also result in groundwater pollution with pharmaceutical residues derived from veterinary drugs.

The US Environmental Protection Agency (EPA) and the European Commission are seriously dealing with the nitrate problem related to agricultural development, as a major water supply problem that requires appropriate management and governance.

Runoff of pesticides may leach into groundwater causing human health problems from contaminated water wells. Pesticide concentrations found in groundwater are typically low, and often the regulatory human health-based limits exceeded are also very low. The organophosphorus insecticide monocrotophos (MCP) appears to be one of a few hazardous, persistent, soluble and mobile (it does not bind with minerals in soils) pesticides able to reach a drinking-water source. In general, more pesticide compounds are being detected as groundwater quality monitoring programs have become more extensive; however, much less monitoring has been conducted in developing countries due to the high analysis costs.

Commercial and Industrial Leaks

A wide variety of both inorganic and organic pollutants have been found in aquifers underlying commercial and industrial activities.

Ore mining and metal processing facilities are the primary responsible of the presence of metals in groundwater of anthropogenic origin, including arsenic. The low pH associated with acid mine drainage (AMD) contributes to the solubility of potential toxic metals that can eventually enter the groundwater system.

There is a increasing concern over the groundwater pollution by gasoline leaked from petroleum underground storage tanks (USTs) of gas stations. BTEX compounds are the most common additives of the gasoline. BTEX compounds, including benzene, have densities lower than water (1 g/ml). Similar to the oil spills on the sea, the non-miscible phase, referred to as Light Non-Aqueous Phase Liquid (LNAPL), will "float" upon the water table in the aquifer.

Chlorinated solvents are used in nearly any industrial practice where degreasing removers are required. PCE is a highly utilized solvent in the dry cleaning industry because of its cleaning effectiveness and relatively low cost. It has also been used for metal-degreasing operations. Because it is is highly volatile, it is more frequently found in groundwater than in surface water. TCE has historically been used as a metal cleaning. The military facility Anniston Army Dept (ANAD) in the United States was placed on the US EPA Superfund National Priorities List (NPL) because of groundwater contamination with as much as 27 million pounds of TCE. Both PCE and TCE may degrade to vinyl chloride (VC), the most toxic chlorinated hydrocarbon.

Many types of solvents may have also been disposed illegally, leaking over time to the groundwater system.

Chlorinated solvents such as PCE and TCE have densities higher than water and the non-miscible phase is referred to as Dense Non-Aqueous Phase Liquids (DNAPL). Once they reach the aquifer, they will "sink" and eventually accumulate on the top of low-permeability layers.

Historically, wood-treating facilities have also release insecticides such as pentachlorophenol (PCP) and creosote into the environment, impacting the groundwater resources. PCP is a a highly soluble and toxic obsolete pesticide recently listed in the Stockholm Convention on Persistent Organic Pollutants. PAHs and other semi-VOCs are the common contaminants associated with creosote.

Although non-miscible, both LNAPLs and DNAPLs still have the potential to slowly dissolve into the aqueous (miscible) phase to create a plume and thus become a long-term source of contamination. DNAPLs (chlorinated solvents, heavy PAHs, creosote, PCBs) tend to be difficult to manage as they can reside very deep in the groundwater system.

Hydraulic Fracturing

The recent growth of Hydraulic Fracturing ("Fracking") wells in the United States has raised concerns regarding its potential risks of contaminating groundwater resources. The Environmental

Protection Agency (EPA), along with many other researchers, has been delegated to study the relationship between hydraulic fracturing and drinking water resources. While it is possible to perform hydraulic fracturing without having a relevant impact on groundwater resources if stringent controls und quality management measures are in place, there are a number of cases where groundwater pollution due to improper handling or technical failures was observed.

While the EPA has not found significant evidence of a widespread, systematic impact on drinking water by hydraulic fracturing, this may be due to insufficient systematic pre- and post- hydraulic fracturing data on drinking water quality, and the presence of other agents of contamination that preclude the link between shale oil/gas extraction and its impact.

Despite the EPA's lack of profound widespread evidence, other researchers have made significant observations of rising groundwater contamination in close proximity to major shale oil/gas drilling sites located in Marcellus *Ellsworth, William (2013). "Injection-Induced Earthquakes". Science* (British Columbia, Canada). Within one kilometer of these specific sites, a subset of shallow drinking water consistently showed higher concentration levels of methane, ethane, and propane concentrations than normal. An evaluation of higher Helium and other noble gas concentration along with the rise of hydrocarbon levels supports the distinction between hydraulic fracturing fugitive gas and naturally occurring "background" hydrocarbon content. This contamination is speculated to be the result of leaky, failing, or improperly installed gas well casings.

Furthermore, it is theorized that contamination could also result from the capillary migration of deep residual hyper-saline water and hydraulic fracturing fluid, slowly flowing through faults and fractures until finally making contact with groundwater resources; however, many researchers argue that the permeability of rocks overlying shale formations are too low to allow this to ever happen sufficiently. To ultimately prove this theory, there would have to be traces of toxic trihalomethanes (THM) since they are often associated with the presence of stray gas contamination, and typically co-occur with high halogen concentrations in hyper-saline waters. Besides, highly saline waters are a common natural feature in deep groundwater systems.

While conclusions regarding groundwater pollution as the result to hydraulic fracturing fluid flow is restricted in both space and time, researchers have hypothesized that the potential for systematic stray gas contamination depends mainly on the integrity of the shale oil/gas well structure, along with its relative geological location to local fracture systems that could potentially provide flow paths for fugitive gas migration.

Though widespread, systematic contamination by hydraulic fracturing has been heavily disputed, one major source of contamination that has the most consensus among researchers of being the most problematic is site-specific accidental spillage of hydraulic fracturing fluid and produced water. So far, a significant majority of groundwater contamination events are derived from surface-level anthropogenic routes rather than the subsurface flow from underlying shale formations. Examples of such events include: a fracking fluid spillage in Acorn Fork Creek, Kentucky that caused a widespread death among aquatic species in 2007; a 420,000 gallon spillage of hyper-saline produced water that turned a once very-fertile farmland in New Mexico into a dead-zone in 2010; and a 42,000 gallon fracking fluid spillage in Arlington, Texas that necessitated an evacuation of over a 100 homes in 2015. While the damage can be obvious, and much more effort is being done to prevent these accidents from occurring so frequently, the lack of data from fracking oil spills continue to leave researchers in the dark. In many of these events, the data acquired from the leakage or spillage is often very vague, and thus would lead researchers to lacking conclusions.

Researchers from the Federal Institute for Geosciences and Natural Resources (BGR) conducted a modelling study for a deep shale-gas formation in the North German Basin. They concluded that the probability is small that the rise of fracking fluids through the geological underground to the surface will impact shallow groundwater.

Landfill Leachate

Leachate from sanitary landfills can lead to groundwater pollution.

Love Canal was one of the most widely known examples of groundwater pollution. In 1978, residents of the Love Canal neighborhood in upstate New York noticed high rates of cancer and an alarming number of birth defects. This was eventually traced to organic solvents and dioxins from an industrial landfill that the neighborhood had been built over and around, which had then infiltrated into the water supply and evaporated in basements to further contaminate the air. Eight hundred families were reimbursed for their homes and moved, after extensive legal battles and media coverage.

Other

Further causes of groundwater pollution are chemical spills from commercial or industrial operations, chemical spills occurring during transport (e.g. spillage of diesel fuels), illegal waste dumping, infiltration from urban runoff or mining operations, road salts, de-icing chemicals from airports and even atmospheric contaminants since groundwater is part of the hydrologic cycle.

The burial of corpses and their subsequent degradation may also pose a risk of pollution to groundwater.

Mechanisms

The passage of water through the subsurface can provide a reliable natural barrier to contamination but it only works under favorable conditions.

The stratigraphy of the area plays an important role in the transport of pollutants. An area can have layers of sandy soil, fractured bedrock, clay, or hardpan. Areas of karst topography on limestone bedrock are sometimes vulnerable to surface pollution from groundwater. Earthquake faults can also be entry routes for downward contaminant entry. Water table conditions are of great importance for drinking water supplies, agricultural irrigation, waste disposal (including nuclear waste), wildlife habitat, and other ecological issues.

Interactions with Surface Water

Although interrelated, surface water and groundwater have often been studied and managed as separate resources. Surface water seeps through the soil and becomes groundwater. Conversely, groundwater can also feed surface water sources. Sources of surface water pollution are generally grouped into two categories based on their origin.

Interactions between groundwater and surface water are complex. Consequently, groundwater pollution, sometimes referred to as groundwater contamination, is not as easily classified as surface water pollution. By its very nature, groundwater aquifers are susceptible to contamination from sources that may not directly affect surface water bodies, and the distinction of point vs. non-point source may be irrelevant. A spill or ongoing release of chemical or radionuclide contaminants into soil (located away from a surface water body) may not create point or non-point source pollution but can contaminate the aquifer below, creating a toxic plume.

Prevention

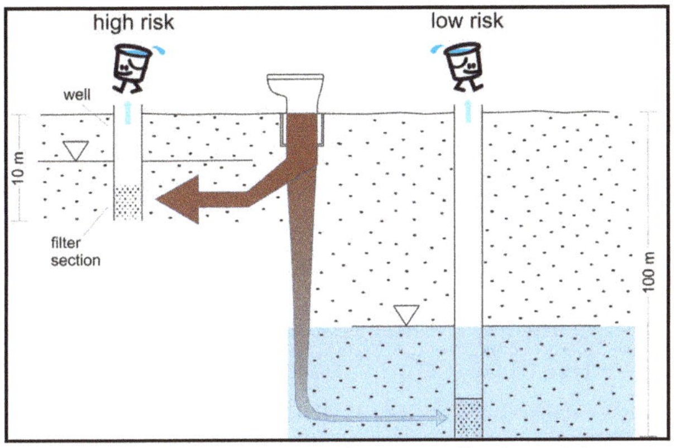

Schematic showing that there is a lower risk of groundwater pollution with greater depth of the water well.

Precautionary Principle

The precautionary principle, evolved from Principle 15 of the Rio Declaration on Environment and Development, is important in protecting groundwater resources from pollution. The precautionary principle provides that *"where there are threats of irreversible damage, lack of full scientific certainty shall not be used as reason for postponing cost-effective measures to prevent environmental degradation."*.

One of the six basic principles of the European Union (EU) water policy is the application of the precautionary principle.

Groundwater Quality Monitoring

Groundwater quality monitoring programs have been implemented regularly in many countries around the world. They are important components to understand the hydrogeological system, and for the development of conceptual models and aquifer vulnerability maps.

Groundwater quality must be regularly monitored across the aquifer to determine trends. Effective groundwater monitoring should be driven by a specific objective, for example, a specific contaminant of concern. Contaminant levels can be compared to the World Health Organization (WHO) guidelines for drinking-water quality. It is not rare that limits of contaminants are reduced as more medical experience is gained.

Sufficient investment should be given to continue monitoring over the long term. When a problem is found, action should be taken to correct it. Waterborne outbreaks in the United States decreased with the introduction of more stringent monitoring (and treatment) requirements in the early 90s.

The community can also help monitor the groundwater quality.

Land Zoning for Groundwater Protection

The development of land-use zoning maps has been implemented by several water authorities at different scales around the world. There are two types of zoning maps: aquifer vulnerability maps and source protection maps.

Aquifer Vulnerability Map

It refers to the intrinsic (or natural) vulnerability of a groundwater system to pollution. Intrinsically, some aquifers are more vulnerable to pollution than other aquifers. Shallow unconfined aquifers are more at risk of pollution because there are fewer layers to filter out contaminants.

The unsaturated zone can play an important role in retarding (and in some cases eliminating) pathogens and so must be considered when assessing aquifer vulnerability. The biological activity is greatest in the top soil layers where the attenuation of pathogens is generally most effective.

Preparation of the vulnerability maps typically involves overlaying several thematic maps of physical factors that have been selected to describe the aquifer vulnerability. The index-based parametric mapping method GOD developed by Foster and Hirata (1988) uses three generally available or readily estimated parameters, the degree of Groundwater hydraulic confinement, geological nature of the Overlying strata and Depth to groundwater. A further approach developed by the US EPA named DRASTIC employs seven hydrogeological factors to develop an index of vulnerability: Depth to water table, net Recharge, Aquifer media, Soil media, Topography (slope), Impact on the vadose zone, and hydraulic Conductivity.

There is a particular debate among hydrogeologist whether aquifer vulnerability should be established in a general (intrinsic) way for all contaminants, or specifically for each pollutant.

Source Protection Map

It refers to the capture areas around an individual groundwater source, such as a water well or a spring, to especially protect them from pollution. Thus, potential sources of degradable pollutants, such as pathogens, can be located at distances which travel times along the flowpaths are long enough for the pollutant to be eliminated through filtration or adsorption.

Analytical methods using equations to define groundwater flow and contaminant transport are the most widely used. The WHPA is a semi-analytical groundwater flow simulation program developed by the US EPA for delineating capture zones in a wellhead protection area.

The simplest form of zoning employs fixed-distance methods where activities are excluded within a uniformly applied specified distance around abstraction points.

Locating On-site Sanitation Systems

As the health effects of most toxic chemicals arise after prolonged exposure, risk to health from chemicals is generally lower than that from pathogens. Thus, the quality of the source protection measures is an important component in controlling whether pathogens may be present in the final drinking-water.

On-site sanitation systems can be designed in such a way that groundwater pollution from these sanitation systems is prevented from occurring. Detailed guidelines have been developed to estimate safe distances to protect groundwater sources from pollution from on-site sanitation. The following criteria have been proposed for safe siting (i.e. deciding on the location) of on-site sanitation systems:

- Horizontal distance between the drinking water source and the sanitation system:
    - Guideline values for horizontal separation distances between on-site sanitation systems and water sources vary widely (e.g. 15 to 100 m horizontal distance between pit latrine and groundwater wells).
- Vertical distance between drinking water well and sanitation system.
- Aquifer type.
- Groundwater flow direction.
- Impermeable layers.
- Slope and surface drainage.
- Volume of leaking wastewater.
- Superposition, i.e. the need to consider a larger planning area.

As a very general guideline it is recommended that the bottom of the pit should be at least 2 m above groundwater level, and a minimum horizontal distance of 30 m between a pit and a water source is normally recommended to limit exposure to microbial contamination. However, no general statement should be made regarding the minimum lateral separation distances required to prevent contamination of a well from a pit latrine. For example, even 50 m lateral separation distance might not be sufficient in a strongly karstified system with a downgradient supply well or spring, while 10 m lateral separation distance is completely sufficient if there is a well developed clay cover layer and the annular space of the groundwater well is well sealed.

Legislation

Sign near Mannheim, Germany indicating a zone as a dedicated "groundwater protection zone".

Institutional and legal issues are critical in determining the success or failure of groundwater protection policies and strategies.

United States

In November 2006, the Environmental Protection Agency published the Ground Water Rule in the United States Federal Register. The EPA was worried that the ground water system would be vulnerable to contamination from fecal matter. The point of the rule was to keep microbial pathogens out of public water sources. The 2006 Ground Water Rule was an amendment of the 1996 Safe Drinking Water Act.

The ways to deal with groundwater pollution that has already occurred can be grouped into the following categories: containing the pollutants to prevent them from migrating further; removing the pollutants from the aquifer; remediating the aquifer by either immobilizing or detoxifying the contaminants while they are still in the aquifer (in-situ); treating the groundwater at its point of use; or abandoning the use of this aquifer's groundwater and finding an alternative source of water.

Management

Point-of-use Treatment

Portable water purification devices or "point-of-use" (POU) water treatment systems and field water disinfection techniques can be used to remove some forms of groundwater pollution prior to drinking, namely any fecal pollution. Many commercial portable water purification systems or chemical additives are available which can remove pathogens, chlorine, bad taste, odors, and heavy metals like lead and mercury.

Techniques include boiling, filtration, activated charcoal absorption, chemical disinfection, ultraviolet purification, ozone water disinfection, solar water disinfection, solar distillation, homemade water filters.

Arsenic removal filters (ARF) are dedicated technologies typically installed to remove arsenic. Many of these technologies require a capital investment and long-term maintenance. Filters in Bangladesh are usually abandoned by the users due to their high cost and complicated maintenance, which is also quite expensive.

Groundwater Remediation

Groundwater pollution is much more difficult to abate than surface pollution because groundwater can move great distances through unseen aquifers. Non-porous aquifers such as clays partially purify water of bacteria by simple filtration (adsorption and absorption), dilution, and, in some cases, chemical reactions and biological activity; however, in some cases, the pollutants merely transform to soil contaminants. Groundwater that moves through open fractures and caverns is not filtered and can be transported as easily as surface water. In fact, this can be aggravated by the human tendency to use natural sinkholes as dumps in areas of karst topography.

Pollutants and contaminants can be removed from ground water by applying various techniques thereby making it safe for use. Ground water treatment (or remediation) techniques span biological, chemical, and physical treatment technologies. Most ground water treatment techniques utilize a combination of technologies. Some of the biological treatment techniques include bioaugmentation, bioventing, biosparging, bioslurping, and phytoremediation. Some chemical treatment techniques include ozone and oxygen gas injection, chemical precipitation, membrane separation, ion exchange, carbon absorption, aqueous chemical oxidation, and surfactant enhanced recovery. Some chemical techniques may be implemented using nanomaterials. Physical treatment techniques include, but are not limited to, pump and treat, air sparging, and dual phase extraction.

Abandonment

If treatment or remediation of the polluted groundwater is deemed to be too difficult or expensive then abandoning the use of this aquifer's groundwater and finding an alternative source of water is the only other option.

Thermal Pollution

The Brayton Point Power Station in Massachusetts discharges heated water to Mount Hope Bay.

Thermal pollution is the degradation of water quality by any process that changes ambient water temperature. A common cause of thermal pollution is the use of water as a coolant by power plants and industrial manufacturers. When water used as a coolant is returned to the natural environment at a higher temperature, the change in temperature decreases oxygen supply and affects ecosystem composition. Fish and other organisms adapted to particular temperature range can be killed by an abrupt change in water temperature (either a rapid increase or decrease) known as "thermal shock."

Urban runoff—stormwater discharged to surface waters from roads and parking lots—can also be a source of elevated water temperatures.

Ecological Effects

Potrero Generating Station discharged heated water into San Francisco Bay. The plant was closed in 2011.

Warm Water

Elevated temperature typically decreases the level of dissolved oxygen of water, as gases are less soluble in hotter liquids. This can harm aquatic animals such as fish, amphibians and other aquatic organisms. Thermal pollution may also increase the metabolic rate of aquatic animals, as enzyme activity, resulting in these organisms consuming more food in a shorter time than if their environment were not changed. An increased metabolic rate may result in fewer resources; the more adapted organisms moving in may have an advantage over organisms that are not used to the warmer temperature. As a result, food chains of the old and new environments may be compromised. Some fish species will avoid stream segments or coastal areas adjacent to a thermal discharge. Biodiversity can be decreased as a result.

High temperature limits oxygen dispersion into deeper waters, contributing to anaerobic conditions. This can lead to increased bacteria levels when there is ample food supply. Many aquatic species will fail to reproduce at elevated temperatures.

Primary producers (e.g. plants, cyanobacteria) are affected by warm water because higher water temperature increases plant growth rates, resulting in a shorter lifespan and species overpopulation. This can cause an algae bloom which reduces oxygen levels.

Temperature changes of even one to two degrees Celsius can cause significant changes in organism metabolism and other adverse cellular biology effects. Principal adverse changes can include rendering cell walls less permeable to necessary osmosis, coagulation of cell proteins, and alteration of enzyme metabolism. These cellular level effects can adversely affect mortality and reproduction.

A large increase in temperature can lead to the denaturing of life-supporting enzymes by breaking down hydrogen- and disulphide bonds within the quaternary structure of the enzymes. Decreased enzyme activity in aquatic organisms can cause problems such as the inability to break down lipids, which leads to malnutrition.

In limited cases, warm water has little deleterious effect and may even lead to improved function of the receiving aquatic ecosystem. This phenomenon is seen especially in seasonal waters and is known as thermal enrichment. An extreme case is derived from the aggregational habits of the manatee, which often uses power plant discharge sites during winter. Projections suggest that manatee populations would decline upon the removal of these discharges.

Cold Water

Releases of unnaturally cold water from reservoirs can dramatically change the fish and macroinvertebrate fauna of rivers, and reduce river productivity. In Australia, where many rivers have warmer temperature regimes, native fish species have been eliminated, and macroinvertebrate fauna have been drastically altered. This may be mitigated by designing the dam to release warmer surface waters instead of the colder water at the bottom of the reservoir.

Thermal Shock

When a power plant first opens or shuts down for repair or other causes, fish and other organisms adapted to particular temperature range can be killed by the abrupt change in water temperature, either an increase or decrease, known as "thermal shock.":

Sources and Control of Thermal Pollution

Cooling tower at Gustav Knepper Power Station, Dortmund, Germany.

Industrial Wastewater

In the United States, about 75 to 82 percent of thermal pollution is generated by power plants. The remainder is from industrial sources such as petroleum refineries, pulp and paper mills, chemical plants, steel mills and smelters. Heated water from these sources may be controlled with:

- cooling ponds, man-made bodies of water designed for cooling by evaporation, convection, and radiation.
- cooling towers, which transfer waste heat to the atmosphere through evaporation and/or heat transfer.
- cogeneration, a process where waste heat is recycled for domestic and/or industrial heating purposes.

Some facilities use once-through cooling (OTC) systems which do not reduce temperature as effectively as the above systems. For example, the Potrero Generating Station in San Francisco (closed in 2011), used OTC and discharged water to San Francisco Bay approximately 10 °C (20 °F) above the ambient bay temperature.

A bioretention cell for treating urban runoff in California.

Urban Runoff

During warm weather, urban runoff can have significant thermal impacts on small streams, as storm water passes over hot parking lots, roads and sidewalks. Storm water management facilities that absorb runoff or direct it into groundwater, such as bioretention systems and infiltration basins, can reduce these thermal effects. These and related systems for managing runoff are components of an expanding urban design approach commonly called green infrastructure.

Retention basins (stormwater ponds) tend to be less effective at reducing runoff temperature, as the water may be heated by the sun before being discharged to a receiving stream.

Marine Pollution

While marine pollution can be obvious, as with the marine debris shown above, it is often the pollutants that cannot be seen that cause most harm.

Marine pollution occurs when harmful, or potentially harmful, effects result from the entry into the ocean of chemicals, particles, industrial, agricultural and residential waste, noise, or the spread of invasive organisms. Eighty percent of marine pollution comes from land. Air pollution is also a contributing factor by carrying off pesticides or dirt into the ocean. Land and air pollution have proven to be harmful to marine life and its habitats.

The pollution often comes from non point sources such as agricultural runoff, wind-blown debris and dust. Nutrient pollution, a form of water pollution, refers to contamination by excessive inputs of nutrients. It is a primary cause of eutrophication of surface waters, in which excess nutrients, usually nitrogen or phosphorus, stimulate algae growth.

Many potentially toxic chemicals adhere to tiny particles which are then taken up by plankton and benthos animals, most of which are either deposit or filter feeders. In this way, the toxins are concentrated upward within ocean food chains. Many particles combine chemically in a manner highly depletive of oxygen, causing estuaries to become anoxic.

When pesticides are incorporated into the marine ecosystem, they quickly become absorbed into marine food webs. Once in the food webs, these pesticides can cause mutations, as well as diseases, which can be harmful to humans as well as the entire food web.

Toxic metals can also be introduced into marine food webs. These can cause a change to tissue matter, biochemistry, behaviour, reproduction, and suppress growth in marine life. Also, many animal feeds have a high fish meal or fish hydrolysate content. In this way, marine toxins can be transferred to land animals, and appear later in meat and dairy products.

History

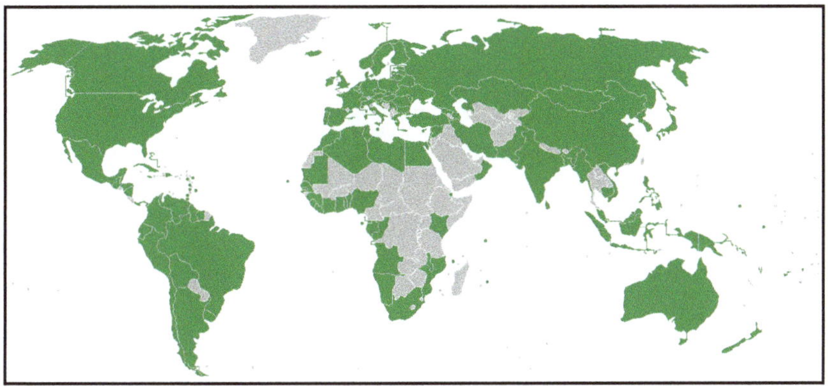

Parties to the MARPOL 73/78 convention on marine pollution.

Although marine pollution has a long history, significant international laws to counter it were only enacted in the twentieth century. Marine pollution was a concern during several United Nations Conferences on the Law of the Sea beginning in the 1950s. Most scientists believed that the oceans were so vast that they had unlimited ability to dilute, and thus render pollution harmless.

In the late 1950s and early 1960s, there were several controversies about dumping radioactive waste off the coasts of the United States by companies licensed by the Atomic Energy Commission, into the Irish Sea from the British reprocessing facility at Windscale, and into the Mediterranean Sea by the French Commissariat à l'Energie Atomique. After the Mediterranean Sea controversy, for example, Jacques Cousteau became a worldwide figure in the campaign to stop marine pollution. Marine pollution made further international headlines after the 1967 crash of the oil tanker Torrey Canyon, and after the 1969 Santa Barbara oil spill off the coast of California.

Marine pollution was a major area of discussion during the 1972 United Nations Conference on the Human Environment, held in Stockholm. That year also saw the signing of the Convention on the Prevention of Marine Pollution by Dumping of Wastes and Other Matter, sometimes called the London Convention. The London Convention did not ban marine pollution, but it established black and gray lists for substances to be banned (black) or regulated by national authorities (gray). Cyanide and high-level radioactive waste, for example, were put on the black list. The London Convention applied only to waste dumped from ships, and thus did nothing to regulate waste discharged as liquids from pipelines.

Pathways of Pollution

There are many different ways to categorize, and examine the inputs of pollution into our marine ecosystems. Patin (n.d.) notes that generally there are three main types of inputs of pollution into the ocean: direct discharge of waste into the oceans, runoff into the waters due to rain, and pollutants that are released from the atmosphere.

One common path of entry by contaminants to the sea are rivers. The evaporation of water from oceans exceeds precipitation. The balance is restored by rain over the continents entering rivers and then being returned to the sea. The Hudson in New York State and the Raritan in New Jersey, which empty at the northern and southern ends of Staten Island, are a source of mercury contamination of zooplankton (copepods) in the open ocean. The highest concentration in the filter-feeding copepods is not at the mouths of these rivers but 70 miles south, nearer Atlantic City, because water flows close to the coast. It takes a few days before toxins are taken up by the plankton.

Septic river.

Pollution is often classed as point source or nonpoint source pollution. Point source pollution occurs when there is a single, identifiable, and localized source of the pollution. An example is directly discharging sewage and industrial waste into the ocean. Pollution such as this occurs particularly in developing nations. Nonpoint source pollution occurs when the pollution comes from ill-defined and diffuse sources. These can be difficult to regulate. Agricultural runoff and wind blown debris are prime examples.

Direct Discharge

Acid mine drainage in the Rio Tinto River.

Pollutants enter rivers and the sea directly from urban sewerage and industrial waste discharges, sometimes in the form of hazardous and toxic wastes.

Inland mining for copper, gold. etc., is another source of marine pollution. Most of the pollution is simply soil, which ends up in rivers flowing to the sea. However, some minerals discharged in the course of the mining can cause problems, such as copper, a common industrial pollutant, which can interfere with the life history and development of coral polyps. Mining has a poor environmental track record. For example, according to the United States Environmental Protection Agency, mining has contaminated portions of the headwaters of over 40% of watersheds in the western continental US. Much of this pollution finishes up in the sea.

Land Runoff

Surface runoff from farming, as well as urban runoff and runoff from the construction of roads, buildings, ports, channels, and harbours, can carry soil and particles laden with carbon, nitrogen, phosphorus, and minerals. This nutrient-rich water can cause fleshy algae and phytoplankton to thrive in coastal areas; known as algal blooms, which have the potential to create hypoxic conditions by using all available oxygen.

Polluted runoff from roads and highways can be a significant source of water pollution in coastal areas. About 75% of the toxic chemicals that flow into Puget Sound are carried by stormwater that runs off paved roads and driveways, rooftops, yards and other developed land.

Ship Pollution

A cargo ship pumps ballast water over the side.

Ships can pollute waterways and oceans in many ways. Oil spills can have devastating effects. While being toxic to marine life, polycyclic aromatic hydrocarbons (PAHs), found in crude oil, are very difficult to clean up, and last for years in the sediment and marine environment.

Oil spills are probably the most emotive of marine pollution events. However, while a tanker wreck may result in extensive newspaper headlines, much of the oil in the world's seas comes from other smaller sources, such as tankers discharging ballast water from oil tanks used on return ships, leaking pipelines or engine oil disposed of down sewers.

Discharge of cargo residues from bulk carriers can pollute ports, waterways and oceans. In many instances vessels intentionally discharge illegal wastes despite foreign and domestic regulation prohibiting such actions. It has been estimated that container ships lose over 10,000 containers at sea each year (usually during storms). Ships also create noise pollution that disturbs natural wildlife, and water from ballast tanks can spread harmful algae and other invasive species.

Ballast water taken up at sea and released in port is a major source of unwanted exotic marine life. The invasive freshwater zebra mussels, native to the Black, Caspian and Azov seas, were probably transported to the Great Lakes via ballast water from a transoceanic vessel. Meinesz believes that one of the worst cases of a single invasive species causing harm to an ecosystem can be attributed to a seemingly harmless jellyfish. *Mnemiopsis leidyi*, a species of comb jellyfish that spread so it now inhabits estuaries in many parts of the world. It was first introduced in 1982, and thought to have been transported to the Black Sea in a ship's ballast water. The population of the jellyfish shot up exponentially and, by 1988, it was wreaking havoc upon the local fishing industry. "The anchovy catch fell from 204,000 tons in 1984 to 200 tons in 1993; sprat from 24,600 tons in 1984 to 12,000 tons in 1993; horse mackerel from 4,000 tons in 1984 to zero in 1993." Now that the jellyfish have exhausted the zooplankton, including fish larvae, their numbers have fallen dramatically, yet they continue to maintain a stranglehold on the ecosystem.

Invasive species can take over once occupied areas, facilitate the spread of new diseases, introduce new genetic material, alter underwater seascapes and jeopardize the ability of native species to obtain food. Invasive species are responsible for about $138 billion annually in lost revenue and management costs in the US alone.

Atmospheric Pollution

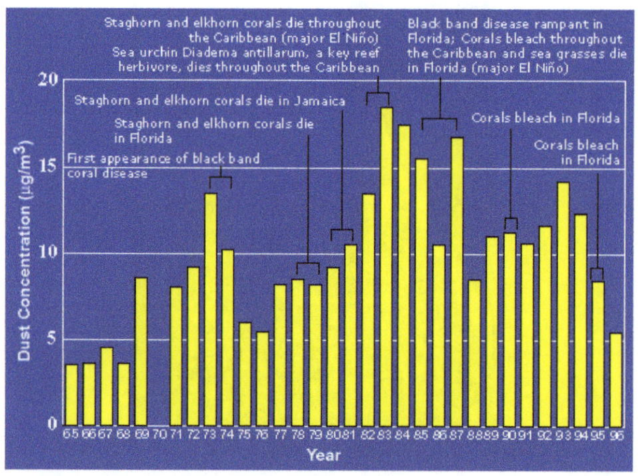

Graph linking atmospheric dust to various coral deaths across the Caribbean Sea and Florida.

Another pathway of pollution occurs through the atmosphere. Wind blown dust and debris, including plastic bags, are blown seaward from landfills and other areas. Dust from the Sahara moving around the southern periphery of the subtropical ridge moves into the Caribbean and Florida during the warm season as the ridge builds and moves northward through the subtropical Atlantic. Dust can also be attributed to a global transport from the Gobi and Taklamakan deserts across Korea, Japan, and the Northern Pacific to the Hawaiian Islands. Since 1970, dust outbreaks have worsened due to periods of drought in Africa. There is a large variability in dust transport to the Caribbean and Florida from year to year; however, the flux is greater during positive phases of the North Atlantic Oscillation. The USGS links dust events to a decline in the health of coral reefs across the Caribbean and Florida, primarily since the 1970s.

Climate change is raising ocean temperatures and raising levels of carbon dioxide in the atmosphere. These rising levels of carbon dioxide are acidifying the oceans. This, in turn, is altering aquatic ecosystems and modifying fish distributions, with impacts on the sustainability of fisheries and the livelihoods of the communities that depend on them. Healthy ocean ecosystems are also important for the mitigation of climate change.

Deep Sea Mining

Deep sea mining is a relatively new mineral retrieval process that takes place on the ocean floor. Ocean mining sites are usually around large areas of polymetallic nodules or active and extinct hydrothermal vents at about 1,400 – 3,700 meters below the ocean's surface. The vents create sulfide deposits, which contain precious metals such as silver, gold, copper, manganese, cobalt, and zinc. The deposits are mined using either hydraulic pumps or bucket systems that take ore to the surface to be processed. As with all mining operations, deep sea mining raises questions about environmental damages to the surrounding areas.

Because deep sea mining is a relatively new field, the complete consequences of full scale mining operations are unknown. However, experts are certain that removal of parts of the sea floor will result in disturbances to the benthic layer, increased toxicity of the water column and sediment plumes from tailings. Removing parts of the sea floor disturbs the habitat of benthic organisms, possibly, depending on the type of mining and location, causing permanent disturbances. Aside from direct impact of mining the area, leakage, spills and corrosion would alter the mining area's chemical makeup.

Among the impacts of deep sea mining, sediment plumes could have the greatest impact. Plumes are caused when the tailings from mining (usually fine particles) are dumped back into the ocean, creating a cloud of particles floating in the water. Two types of plumes occur: near bottom plumes and surface plumes. Near bottom plumes occur when the tailings are pumped back down to the mining site. The floating particles increase the turbidity, or cloudiness, of the water, clogging filter-feeding apparatuses used by benthic organisms. Surface plumes cause a more serious problem. Depending on the size of the particles and water currents the plumes could spread over vast areas. The plumes could impact zooplankton and light penetration, in turn affecting the food web of the area.

Types of Pollution

Acidification

Island with fringing reef in the Maldives. Coral reefs are dying around the world.

The oceans are normally a natural carbon sink, absorbing carbon dioxide from the atmosphere. Because the levels of atmospheric carbon dioxide are increasing, the oceans are becoming more acidic. The potential consequences of ocean acidification are not fully understood, but there are concerns that structures made of calcium carbonate may become vulnerable to dissolution, affecting corals and the ability of shellfish to form shells.

Oceans and coastal ecosystems play an important role in the global carbon cycle and have removed about 25% of the carbon dioxide emitted by human activities between 2000 and 2007 and about half the anthropogenic CO_2 released since the start of the industrial revolution. Rising ocean temperatures and ocean acidification means that the capacity of the ocean carbon sink will gradually get weaker, giving rise to global concerns expressed in the Monaco and Manado Declarations.

A report from NOAA scientists published in the journal Science in May 2008 found that large amounts of relatively acidified water are upwelling to within four miles of the Pacific continental shelf area of North America. This area is a critical zone where most local marine life lives or is born. While the paper dealt only with areas from Vancouver to northern California, other continental shelf areas may be experiencing similar effects.

A related issue is the methane clathrate reservoirs found under sediments on the ocean floors. These trap large amounts of the greenhouse gas methane, which ocean warming has the potential to release. In 2004 the global inventory of ocean methane clathrates was estimated to occupy between one and five million cubic kilometres. If all these clathrates were to be spread uniformly across the ocean floor, this would translate to a thickness between three and fourteen metres. This estimate corresponds to 500–2500 gigatonnes carbon (Gt C), and can be compared with the 5000 Gt C estimated for all other fossil fuel reserves.

Eutrophication

Polluted lagoon.

Effect of eutrophication on marine benthic life.

Eutrophication is an increase in chemical nutrients, typically compounds containing nitrogen or phosphorus, in an ecosystem. It can result in an increase in the ecosystem's primary productivity (excessive plant growth and decay), and further effects including lack of oxygen and severe reductions in water quality, fish, and other animal populations.

The biggest culprit are rivers that empty into the ocean, and with it the many chemicals used as fertilizers in agriculture as well as waste from livestock and humans. An excess of oxygen depleting chemicals in the water can lead to hypoxia and the creation of a dead zone.

Estuaries tend to be naturally eutrophic because land-derived nutrients are concentrated where runoff enters the marine environment in a confined channel. The World Resources Institute has identified 375 hypoxic coastal zones around the world, concentrated in coastal areas in Western Europe, the Eastern and Southern coasts of the US, and East Asia, particularly in Japan. In the ocean, there are frequent red tide algae blooms that kill fish and marine mammals and cause respiratory problems in humans and some domestic animals when the blooms reach close to shore.

In addition to land runoff, atmospheric anthropogenic fixed nitrogen can enter the open ocean. A study in 2008 found that this could account for around one third of the ocean's external (non-recycled) nitrogen supply and up to three per cent of the annual new marine biological production. It has been suggested that accumulating reactive nitrogen in the environment may have consequences as serious as putting carbon dioxide in the atmosphere.

One proposed solution to eutrophication in estuaries is to restore shellfish populations, such as oysters. Oyster reefs remove nitrogen from the water column and filter out suspended solids, subsequently reducing the likelihood or extent of harmful algal blooms or anoxic conditions. Filter feeding activity is considered beneficial to water quality by controlling phytoplankton density and sequestering nutrients, which can be removed from the system through shellfish harvest, buried in the sediments, or lost through denitrification. Foundational work toward the idea of improving marine water quality through shellfish cultivation to was conducted by Odd Lindahl et al., using mussels in Sweden.

Plastic Debris

A mute swan builds a nest using plastic garbage.

Marine debris is mainly discarded human rubbish which floats on, or is suspended in the ocean. Eighty percent of marine debris is plastic – a component that has been rapidly accumulating since the end of World War II. The mass of plastic in the oceans may be as high as 100,000,000 tonnes (98,000,000 long tons; 110,000,000 short tons).

Discarded plastic bags, six pack rings and other forms of plastic waste which finish up in the ocean present dangers to wildlife and fisheries. Aquatic life can be threatened through entanglement, suffocation, and ingestion. Fishing nets, usually made of plastic, can be left or lost in the ocean by fishermen. Known as ghost nets, these entangle fish, dolphins, sea turtles, sharks, dugongs, crocodiles, seabirds, crabs, and other creatures, restricting movement, causing starvation, laceration and infection, and, in those that need to return to the surface to breathe, suffocation.

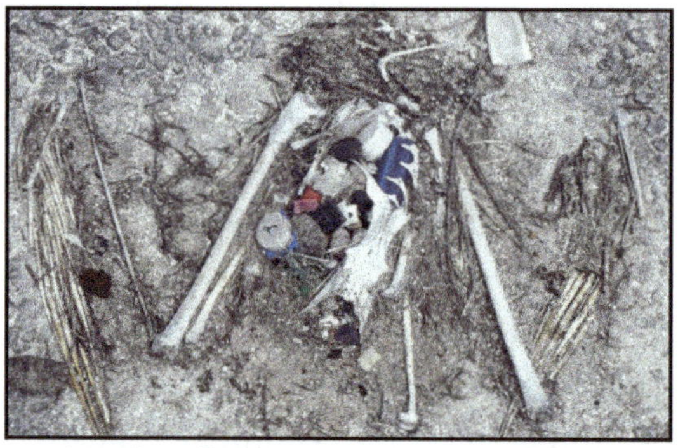

Remains of an albatross containing ingested flotsam.

Many animals that live on or in the sea consume flotsam by mistake, as it often looks similar to their natural prey. Plastic debris, when bulky or tangled, is difficult to pass, and may become permanently lodged in the digestive tracts of these animals. Especially when evolutionary adaptions make it impossible for the likes of turtles to reject plastic bags, which resemble jellyfish when immersed in water, as they have a system in their throat to stop slippery foods from otherwise escaping. Thereby blocking the passage of food and causing death through starvation or infection.

Plastics accumulate because they don't biodegrade in the way many other substances do. They will photodegrade on exposure to the sun, but they do so properly only under dry conditions, and water inhibits this process. In marine environments, photodegraded plastic disintegrates into ever smaller pieces while remaining polymers, even down to the molecular level. When floating plastic particles photodegrade down to zooplankton sizes, jellyfish attempt to consume them, and in this way the plastic enters the ocean food chain. Many of these long-lasting pieces end up in the stomachs of marine birds and animals, including sea turtles, and black-footed albatross.

Marine debris on Kamilo Beach, Hawaii, washed up from the Great Pacific Garbage Patch.

Plastic debris tends to accumulate at the centre of ocean gyres. In particular, the Great Pacific Garbage Patch has a very high level of plastic particulate suspended in the upper water column. In samples taken in 1999, the mass of plastic exceeded that of zooplankton (the dominant animal life in the area) by a factor of six. Midway Atoll, in common with all the Hawaiian Islands, receives substantial amounts of debris from the garbage patch. Ninety percent plastic, this debris accumulates on the beaches of Midway where it becomes a hazard to the bird population of the island.

Midway Atoll is home to two-thirds (1.5 million) of the global population of Laysan albatross. Nearly all of these albatross have plastic in their digestive system and one-third of their chicks die.

Toxic additives used in the manufacture of plastic materials can leach out into their surroundings when exposed to water. Waterborne hydrophobic pollutants collect and magnify on the surface of plastic debris, thus making plastic far more deadly in the ocean than it would be on land. Hydrophobic contaminants are also known to bioaccumulate in fatty tissues, biomagnifying up the food chain and putting pressure on apex predators. Some plastic additives are known to disrupt the endocrine system when consumed, others can suppress the immune system or decrease reproductive rates. Floating debris can also absorb persistent organic pollutants from seawater, including PCBs, DDT and PAHs. Aside from toxic effects, when ingested some of these are mistaken by the animal brain for estradiol, causing hormone disruption in the affected wildlife.

Toxins

Apart from plastics, there are particular problems with other toxins that do not disintegrate rapidly in the marine environment. Examples of persistent toxins are PCBs, DDT, TBT, pesticides, furans, dioxins, phenols and radioactive waste. Heavy metals are metallic chemical elements that have a relatively high density and are toxic or poisonous at low concentrations. Examples are mercury, lead, nickel, arsenic and cadmium. Such toxins can accumulate in the tissues of many species of aquatic life in a process called bioaccumulation. They are also known to accumulate in benthic environments, such as estuaries and bay muds: a geological record of human activities of the last century.

Specific examples:

- Chinese and Russian industrial pollution such as phenols and heavy metals in the Amur River have devastated fish stocks and damaged its estuary soil.
- Wabamun Lake in Alberta, Canada, once the best whitefish lake in the area, now has unacceptable levels of heavy metals in its sediment and fish.
- Acute and chronic pollution events have been shown to impact southern California kelp forests, though the intensity of the impact seems to depend on both the nature of the contaminants and duration of exposure.
- Due to their high position in the food chain and the subsequent accumulation of heavy metals from their diet, mercury levels can be high in larger species such as bluefin and albacore. As a result, in March 2004 the United States FDA issued guidelines recommending that pregnant women, nursing mothers and children limit their intake of tuna and other types of predatory fish.
- Some shellfish and crabs can survive polluted environments, accumulating heavy metals or toxins in their tissues. For example, mitten crabs have a remarkable ability to survive in highly modified aquatic habitats, including polluted waters. The farming and harvesting of such species needs careful management if they are to be used as a food.

- Surface runoff of pesticides can alter the gender of fish species genetically, transforming male into female fish.
- Heavy metals enter the environment through oil spills – such as the Prestige oil spill on the Galician coast – or from other natural or anthropogenic sources.
- In 2005, the 'Ndrangheta, an Italian mafia syndicate, was accused of sinking at least 30 ships loaded with toxic waste, much of it radioactive. This has led to widespread investigations into radioactive-waste disposal rackets.
- Since the end of World War II, various nations, including the Soviet Union, the United Kingdom, the United States, and Germany, have disposed of chemical weapons in the Baltic Sea, raising concerns of environmental contamination.

Underwater Noise

Marine life can be susceptible to noise or the sound pollution from sources such as passing ships, oil exploration seismic surveys, and naval low-frequency active sonar. Sound travels more rapidly and over larger distances in the sea than in the atmosphere. Marine animals, such as cetaceans, often have weak eyesight, and live in a world largely defined by acoustic information. This applies also to many deeper sea fish, who live in a world of darkness. Between 1950 and 1975, ambient noise at one location in the Pacific Ocean increased by about ten decibels (that is a tenfold increase in intensity).

Noise also makes species communicate louder, which is called the Lombard vocal response. Whale songs are longer when submarine-detectors are on. If creatures don't "speak" loud enough, their voice can be masked by anthropogenic sounds. These unheard voices might be warnings, finding of prey, or preparations of net-bubbling. When one species begins speaking louder, it will mask other species voices, causing the whole ecosystem to eventually speak louder.

According to the oceanographer Sylvia Earle, "Undersea noise pollution is like the death of a thousand cuts. Each sound in itself may not be a matter of critical concern, but taken all together, the noise from shipping, seismic surveys, and military activity is creating a totally different environment than existed even 50 years ago. That high level of noise is bound to have a hard, sweeping impact on life in the sea."

Adaptation and Mitigation

Much anthropogenic pollution ends up in the ocean. The 2011 edition of the United Nations Environment Programme Year Book identifies as the main emerging environmental issues the loss to the oceans of massive amounts of phosphorus, "a valuable fertilizer needed to feed a growing global population", and the impact billions of pieces of plastic waste are having globally on the health of marine environments. Bjorn Jennssen (2003) notes in his article, "Anthropogenic pollution may reduce biodiversity and productivity of marine ecosystems, resulting in reduction and depletion of human marine food resources". There are two ways the overall level of this pollution can be mitigated: either the human population is reduced, or a way is found to reduce the ecological footprint left behind by the average human. If the second way is not adopted, then the first way may be imposed as world ecosystems falter.

The second way is for humans, individually, to pollute less. That requires social and political will, together with a shift in awareness so more people respect the environment and are less disposed to abuse it. At an operational level, regulations, and international government participation is needed. It is often very difficult to regulate marine pollution because pollution spreads over international barriers, thus making regulations hard to create as well as enforce.

Aerosol can polluting a beach.

Without appropriate awareness of marine pollution, the necessary global will to effectively address the issues may prove inadequate. Balanced information on the sources and harmful effects of marine pollution need to become part of general public awareness, and ongoing research is required to fully establish, and keep current, the scope of the issues. As expressed in Daoji and Dag's research, one of the reasons why environmental concern is lacking among the Chinese is because the public awareness is low and therefore should be targeted.

Eutrophication

Effects of Eutrophication

Water Pollution:

- Pollution of water because of addition of poisonous substances into water.
- Types of water pollution.

1. Pollution by Point Sources: Contamination of water bodies by substances released from a single, identifiable source such as a pipe- Most common example is the industrial pollution where a lot many harmful substances are released into water bodies.
2. Pollution by nonPoint Sources: Contamination of water bodies by substances that do not come from a discrete, specified source.
    - Pollution by water from farmlands that mixes in streams is a common example for this.
    - Another kind of pollution is ground water pollution when polluted water at the surface sinks into the earth to form ground water.

Point source pollution-Pollutants mixing in water bodies.

A pattern of pollutants getting mixed in a water body.

Waste water from farms contaminates water bodies.

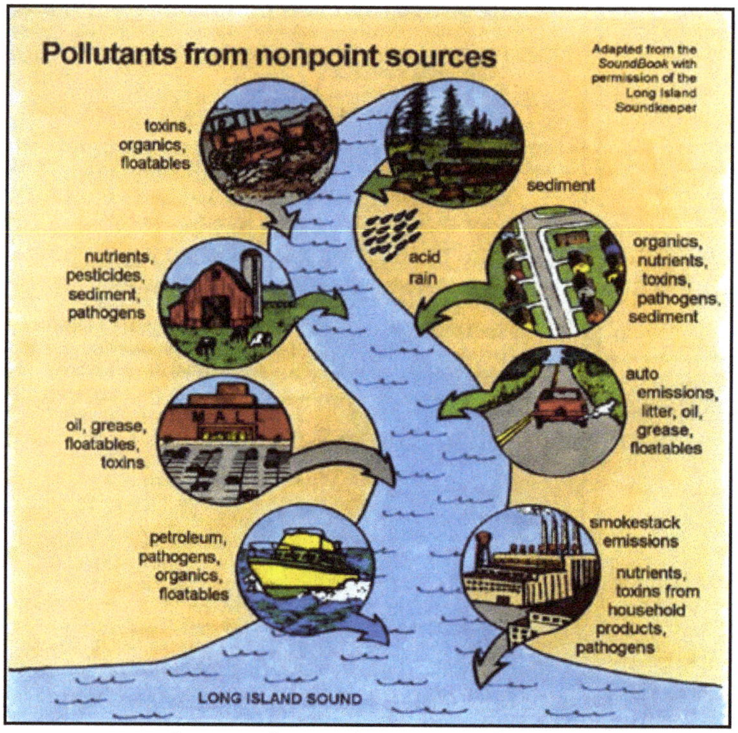

Sources of Nonpoint source pollution.

Eutrophication

- Eutrophication is the process by which a body of water acquires a high concentration of nutrients, especially phosphates and nitrates. These typically promote excessive growth of algae which is called "Algal Bloom". As the algae die and decompose, high levels of organic matter and the decomposing organisms deplete the water of available oxygen, causing the death of other organisms, such as fish.
- Eutrophication can occur naturally. It is considered as a natural ageing process of the water body but human interference has paced up the degradation process.

- The natural process is called "Natural Eutrophication" and Eutrophication by human interference where large amounts of pollutants are released into water because of societies can be termed as "Cultural Eutrophication".

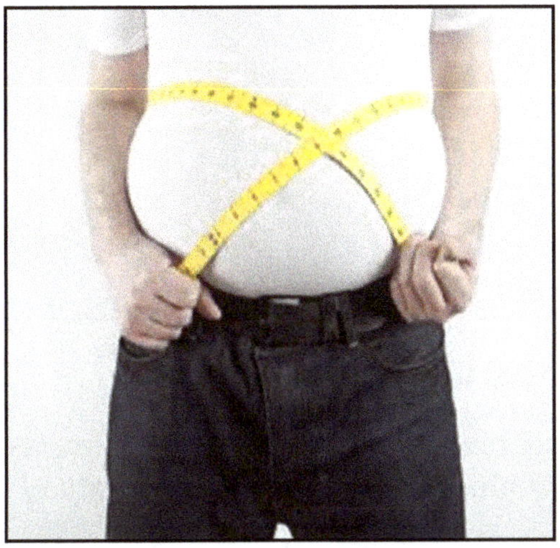

Obesity to humans ~ Eutrophication to water bodies.

Picture showing the excess growth of plants in water because of excess release of nutrients into it.

Eutrophication, naturally is both prominent in fresh water bodies like lakes & rivers and large salt water bodies like seas & oceans.

1. Lakes and rivers generally experience Eutrophication because of increased phosphorous levels and Seas & oceans because of increased Nitrogen levels.
2. Lakes: Classified based on the nutrient content present in them and production of organic matter:
    - Oligotrophic- Less nutrients- In low nutrient medium, life is also less.
    - Mesotrophic - Not many nutrients and not very less nutrients.
    - Eutrophic- High nutrients- some lakes naturally have higher contents of nutrients.
3. In some water bodies, waste released from the living entities present inside them, rain water, atmospheric nitrogen etc., add to the excess nutrients that constitute Eutrophication.

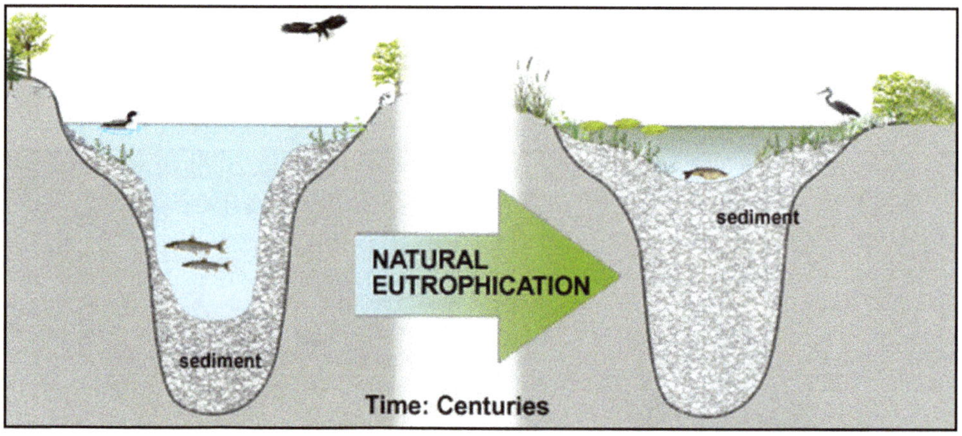

4. Takes a long time for the water body to get polluted to a considerable extent.
5. Factors like availability of light and other resources too influence the growth of unnecessary algae and other organisms that are responsible for Eutrophication-only if it is a long natural process.
6. But because of Human interference and the extent of pollution that is going on now, it would take less than a century for even a large body to get succumbed to Eutrophication.

Cultural Eutrophication

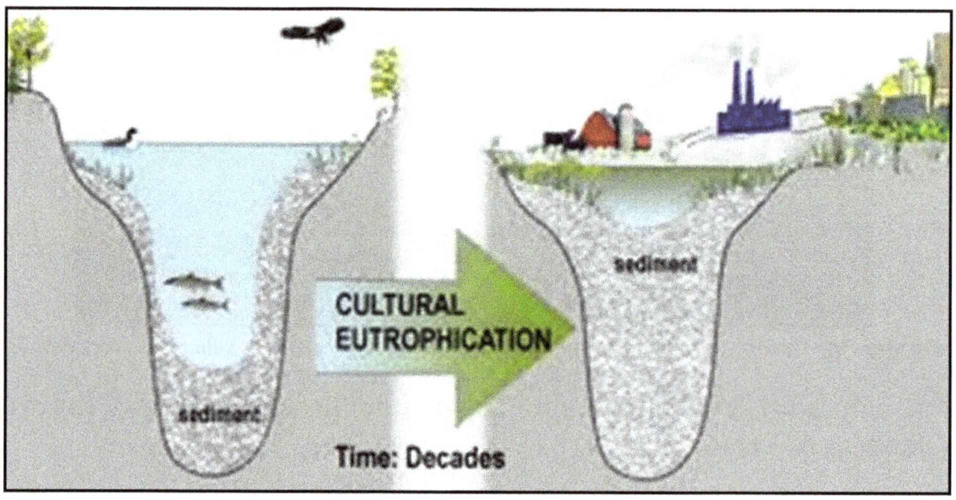

1. Release of pollutants into water which increase the nutrient concentration.
2. Many ways:
    - Sewage disposal in water.
    - Industrial waste disposal in water.
    - Fertilizers and pesticides dissolved in water being channelized into water bodies.
    - Organic fertilizers too can cause Eutrophication.
    - Dissolution of pollutant particles on surface in rain water and flowing into water bodies.

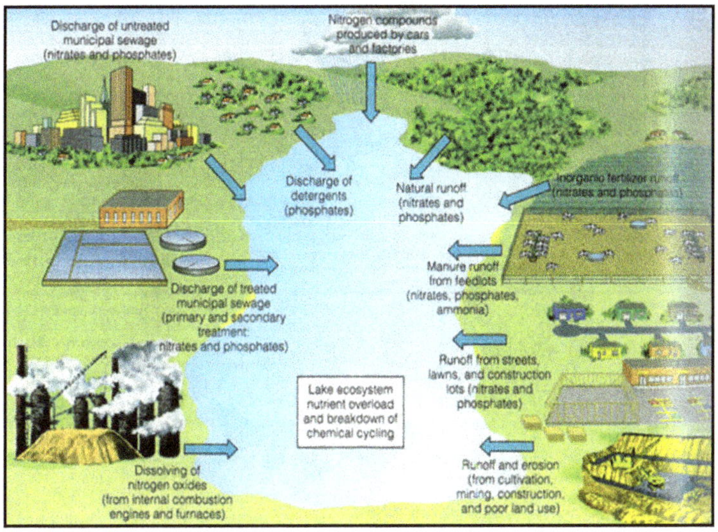

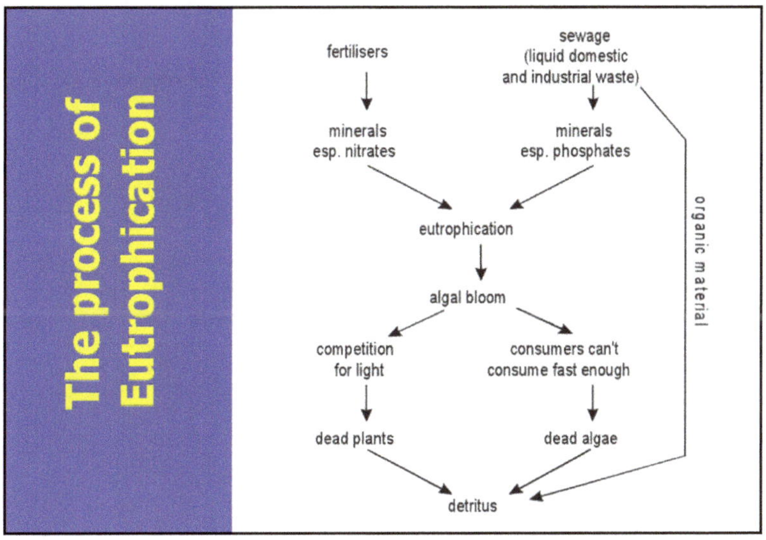

Steps in the Process

1. The process of Eutrophication can be roughly divided into the following steps based on the progress of process:
    - Addition of nutrients.
    - Growth of plants/algae.
    - Death of plants/algae.
    - Growth of bacteria.
    - Lack of Oxygen and other unfavourable conditions for existing living entities.
    - Suffocation and death of living entities, Sedimentation because of excess waste(Detritus).
    - Death of the water body and eventual drying of land.

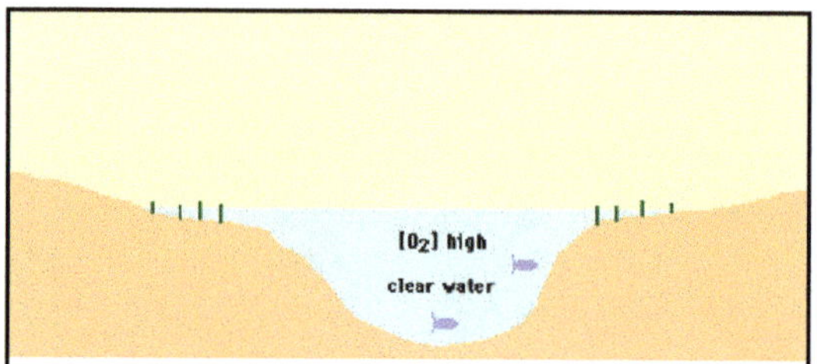

Oligotrophic lake with a low level of nutrients.

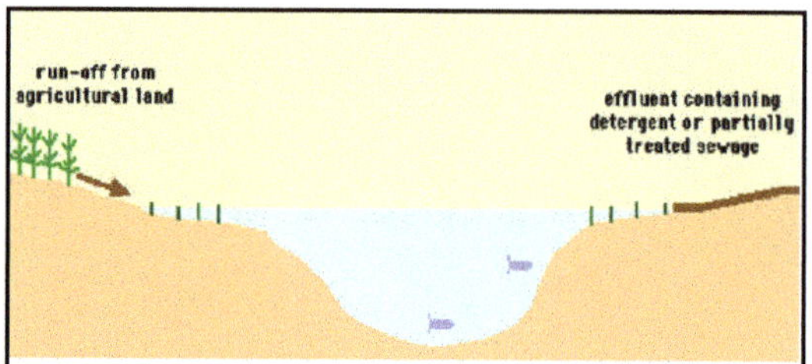

Artificial input of nutrients from run-off and discharge of effluent.

Excess nutrients get added

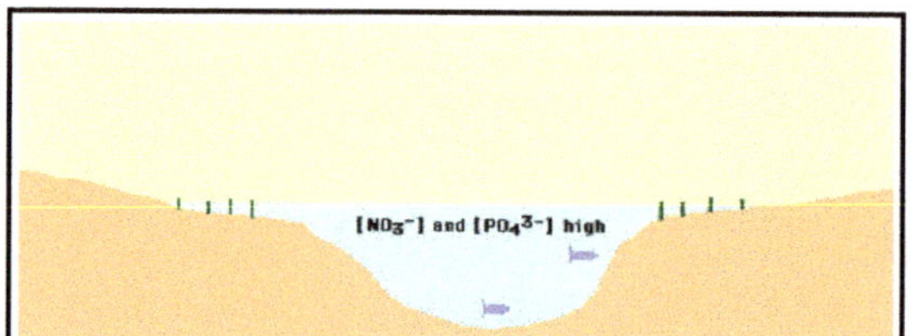

Eutrophic lake with a high level of nutrients. Phosphorus is usually the bio-limiting element in freshwater lakes.

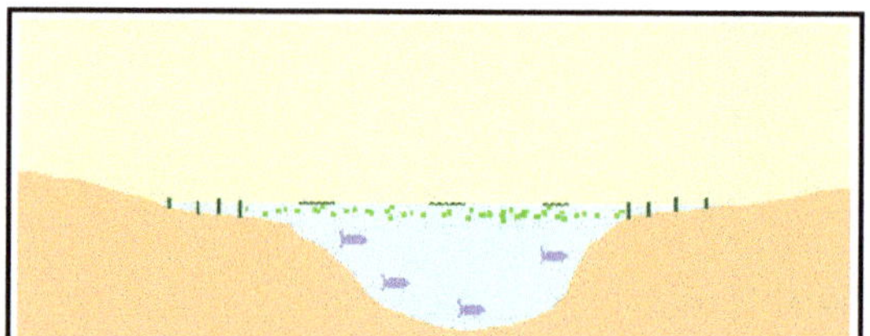

Rapid growth of algae and other biomass resulting in a decrease in the concentration of dissolved oxygen.

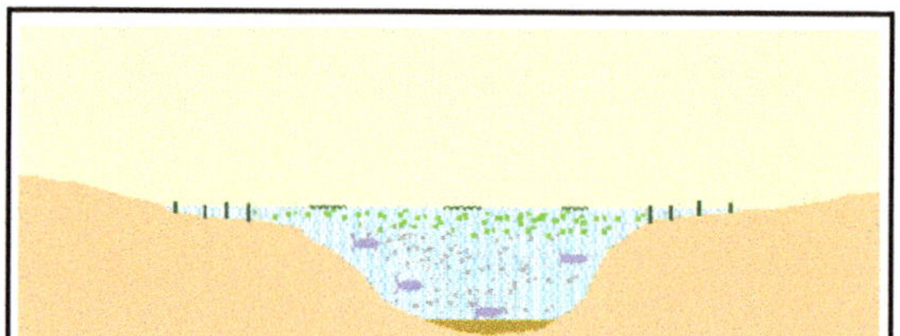

Turbidity (cloudiness) of water increases as does rate of sedimentation.

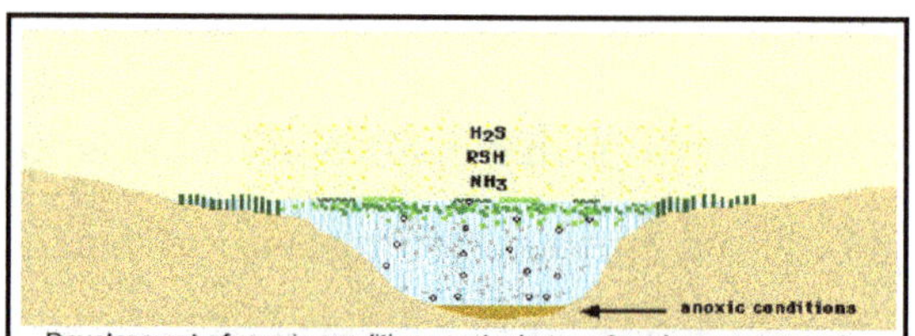

Development of anoxic conditions and release of noxious gases such as hydrogen sulphide, thioalcohols and ammonia.

Effects

- Because of algal bloom, sunlight cannot enter into the depths of the water body which makes plants unable to live ➔ Decrease in O_2 levels occurs resulting in the eventual death of living organisms inside.
- During this process of death of living organisms inside, more oxygen is consumed during the decomposition process.
- Eutrophication not only effects the living organisms inside the water but also degrades the water quality: Water taste and odor degrade because of the toxins like Geosmin , MIB etc., being released during algal bloom.

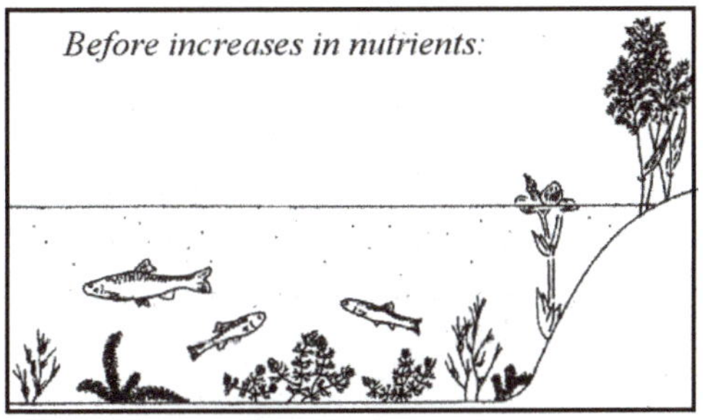

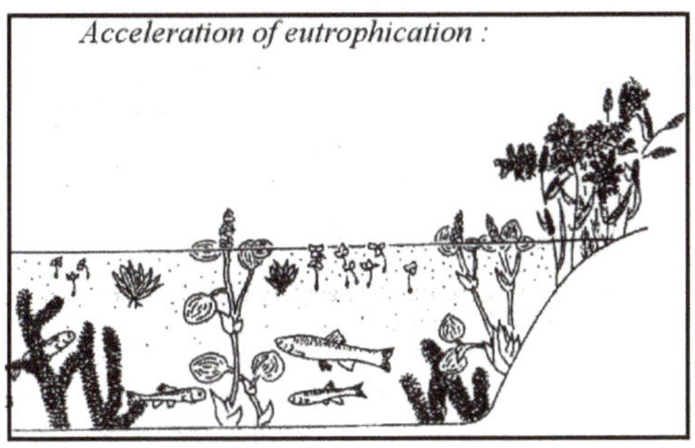

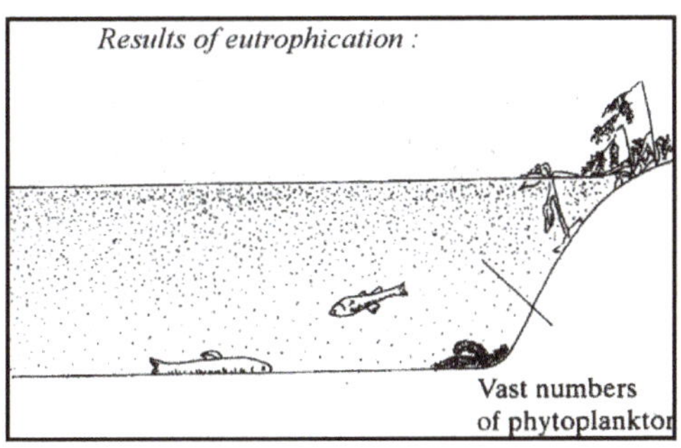

- In lakes and rivers, Eutrophication triggers increase in water level in the long run which creates a problem for shore plant growth and their survival.
- This not only contaminates the water body but eventually ends the plant life on the shore too.

- As water bodies go on dying, the food chain breaks. Only those species which take high nutrients and can grab them fast survive in the habitat.
- Tubificid worms, larvae and bacteria replace the vertebrates in water.
- This broken food chain can lead to extinction of many species and many adverse conditions.

More effects:

- Increased biomass like phytoplankton in water.
- Decrease in harvestable fish and shell fish.
- Increase in turbidity of water.
- Loss of aesthetic value of water body.
- Biological Oxygen Demand(BOD) serves as a measure for finding out about Eutrophication in water.
- Amount of O_2 consumption by living organisms in water- more consumption ➔ more organic matter ➔ more nutrients ➔ Eutrophication.
- A high BOD indicates occurrence of Eutrophication in the water body.

Prevention and Reversal of Eutrophication

- If Organic material addition stops, the bacteria use up all the nutrients available and stop die off after sometime ➔ this can let the left off plants and vertebrates grow their numbers again.
- Concentration of dissolved O_2 increases through natural means like atmospheric replenishment and photosynthesis.
- Cleaning up of water bodies will help in removing sedimentation and allowing plants and fish grow again.
- Organic farming reduces the amount of Nitrites that flow into water when compared to its counterpart of farming.

  And

- Controlling pollution and flow of polluting substances into water bodies.

Acid Mine Drainage

Yellow boy in a stream receiving acid drainage from surface coal mining.

Rocks stained by acid mine drainage on Shamokin Creek.

Acid mine drainage, acid and metalliferous drainage (AMD), or acid rock drainage (ARD) refers to the outflow of acidic water from metal mines or coal mines.

Acid rock drainage occurs naturally within some environments as part of the rock weathering process but is exacerbated by large-scale earth disturbances characteristic of mining and other large construction activities, usually within rocks containing an abundance of sulfide minerals. Areas where the earth has been disturbed (e.g. construction sites, subdivisions, and transportation corridors) may create acid rock drainage. In many localities, the liquid that drains from coal stocks, coal handling facilities, coal washeries, and coal waste tips can be highly acidic, and in such cases it is treated as acid rock drainage.

The same type of chemical reactions and processes may occur through the disturbance of acid sulfate soils formed under coastal or estuarine conditions after the last major sea level rise, and constitutes a similar environmental hazard.

Nomenclature

Historically, the acidic discharges from active or abandoned mines were called acid mine drainage, or AMD. The term acid rock drainage, or ARD, was introduced in the 1980s and 1990s to indicate that acidic drainage can originate from sources other than mines. For example, a paper presented in 1991 at a major international conference on this subject was titled: "The Prediction of Acid Rock Drainage - Lessons from the Database" Both AMD and ARD refer to low pH or acidic waters caused by the oxidation of sulfide minerals, though ARD is the more generic name.

In cases where drainage from a mine is not acidic and has dissolved metals or metalloids, or was originally acidic, but has been neutralized along its flow path, then it is described as "Neutral Mine Drainage", "Mining-Influenced Water" or otherwise. None of these other names have gained general acceptance.

Occurrence

In this case, the pyrite has dissolved away yielding a cube shape and residual gold. This break down is the main driver of acid mine drainage.

Sub-surface mining often progresses below the water table, so water must be constantly pumped out of the mine in order to prevent flooding. When a mine is abandoned, the pumping ceases, and water floods the mine. This introduction of water is the initial step in most acid rock drainage situations. Tailings piles or ponds, mine waste rock dumps, and coal spoils are also an important source of acid mine drainage.

After being exposed to air and water, oxidation of metal sulfides (often pyrite, which is iron-sulfide) within the surrounding rock and overburden generates acidity. Colonies of bacteria and archaea greatly accelerate the decomposition of metal ions, although the reactions also occur in an abiotic environment. These microbes, called extremophiles for their ability to survive in harsh conditions, occur naturally in the rock, but limited water and oxygen supplies usually keep their numbers low. Special extremophiles known as Acidophiles especially favor the low pH levels of abandoned mines. In particular, *Acidithiobacillus ferrooxidans* is a key contributor to pyrite oxidation.

Metal mines may generate highly acidic discharges where the ore is a sulfide mineral or is associated with pyrite. In these cases the predominant metal ion may not be iron but rather zinc, copper, or nickel. The most commonly mined ore of copper, chalcopyrite, is itself a copper-iron-sulfide and occurs with a range of other sulfides. Thus, copper mines are often major culprits of acid mine drainage.

At some mines, acidic drainage is detected within 2–5 years after mining begins, whereas at other mines, it is not detected for several decades. In addition, acidic drainage may be generated for decades or centuries after it is first detected. For this reason, acid mine drainage is considered a serious long-term environmental problem associated with mining.

Chemistry

The chemistry of oxidation of pyrites, the production of ferrous ions and subsequently ferric ions, is very complex, and this complexity has considerably inhibited the design of effective treatment options.

Although a host of chemical processes contribute to acid mine drainage, pyrite oxidation is by far the greatest contributor. A general equation for this process is:

$$2FeS_2(s) + 7O_2(g) + 2H_2O(l) = 2Fe^{2+}(aq) + 4SO_4^{2-}(aq) + 4H^+(aq)$$

The oxidation of the sulfide to sulfate solubilizes the ferrous iron (iron(II)), which is subsequently oxidized to ferric iron (iron(III)):

$$4Fe^{2+}(aq) + O_2(g) + 4H^+(aq) = 4Fe^{3+}(aq) + 2H_2O(l)$$

Either of these reactions can occur spontaneously or can be catalyzed by microorganisms that derive energy from the oxidation reaction. The ferric cations produced can also oxidize additional pyrite and reduce into ferrous ions:

$$FeS_2(s) + 14Fe^{3+}(aq) + 8H_2O(l) = 15Fe^{2+}(aq) + 2SO_4^{2-}(aq) + 16H^+(aq)$$

The net effect of these reactions is to release H^+, which lowers the pH and maintains the solubility of the ferric ion.

Effects

Effects on pH

Water temperatures as high as 47 °C have been measured underground at the Iron Mountain Mine, and the pH can be as low as -3.6.

Organisms which cause acid mine drainage can thrive in waters with pH very close to zero. Negative pH occurs when water evaporates from already acidic pools thereby increasing the concentration of hydrogen ions.

About half of the coal mine discharges in Pennsylvania have pH under 5. However, a significant portion of mine drainage in both the bituminous and anthracite regions of Pennsylvania is alkaline, because limestone in the overburden neutralizes acid before the drainage emanates.

Rio Tinto in Spain.

Acid rock drainage has recently been a hindrance to the completion of the construction of Interstate 99 near State College, Pennsylvania. However, this acid rock drainage didn't come from a mine; rather, it was produced by oxidation of pyrite-rich rock which was unearthed during a road cut and then used as filler material in the I-99 construction. A similar situation developed at the Halifax airport in Canada. It is from these and similar experiences that the term acid *rock* drainage has emerged as being preferable to acid *mine* drainage, thereby emphasizing the general nature of the problem.

Yellow Boy

When the pH of acid mine drainage is raised past 3, either through contact with fresh water or neutralizing minerals, previously soluble iron(III) ions precipitate as iron(III) hydroxide, a yellow-orange solid colloquially known as *yellow boy*. Other types of iron precipitates are possible, including iron oxides and oxyhydroxides. All these precipitates can discolor water and smother plant and animal life on the streambed, disrupting stream ecosystems (a specific offense under the Fisheries Act in Canada). The process also produces additional hydrogen ions, which can further decrease pH. In some cases, the concentrations of iron hydroxides in yellow boy are so high, the precipitate can be recovered for commercial use in pigments.

Trace Metal and Semi-metal Contamination

Many acid rock discharges also contain elevated levels of potentially toxic metals, especially nickel and copper with lower levels of a range of trace and semi-metal ions such as lead, arsenic, aluminium, and manganese. The elevated levels of heavy metals can only be dissolved in waters that have a low pH, as is found in the acidic waters produced by pyrite oxidation. In the coal belt around the south Wales valleys in the UK highly acidic nickel-rich discharges from coal stocking sites have proved to be particularly troublesome.

Identification and Prediction

In a mining setting it is leading practice to carry out a geochemical assessment of mine materials during the early stages of a project to determine the potential for AMD. The geochemical assessment aims to map the distribution and variability of key geochemical parameters, acid generating and element leaching characteristics.

The assessment may include:

1. Sampling;
2. Static geochemical testwork (e.g. acid-base accounting, sulfur speciation);
3. Kinetic geochemical testwork - Conducting oxygen consumption tests, such as the OxCon, to quantify acidity generation rates;
4. Modelling of oxidation, pollutant generation and release;
5. Modelling of material composition.

Treatment

Oversight

In the United Kingdom, many discharges from abandoned mines are exempt from regulatory control. In such cases the Environment Agency working with partners such as the Coal Authority have provided some innovative solutions, including constructed wetland solutions such as on the River Pelenna in the valley of the River Afan near Port Talbot and the constructed wetland next to the River Neath at Ynysarwed.

Although abandoned underground mines produce most of the acid mine drainage, some recently mined and reclaimed surface mines have produced ARD and have degraded local ground-water and surface-water resources. Acidic water produced at active mines must be neutralized to achieve pH 6-9 before discharge from a mine site to a stream is permitted.

In Canada, work to reduce the effects of acid mine drainage is concentrated under the Mine Environment Neutral Drainage (MEND) program. Total liability from acid rock drainage is estimated to be between $2 billion and $5 billion CAD. Over a period of eight years, MEND claims to have reduced ARD liability by up to $400 million CAD, from an investment of $17.5 million CAD.

Methods

Lime Neutralization

By far, the most commonly used commercial process for treating acid mine drainage is lime precipitation in a high-density sludge (HDS) process. In this application, a slurry of lime is dispersed into a tank containing acid mine drainage and recycled sludge to increase water pH to about 9. At this pH, most toxic metals become insoluble and precipitate, aided by the presence of recycled sludge. Optionally, air may be introduced in this tank to oxidize iron and manganese and assist in their precipitation. The resulting slurry is directed to a sludge-settling vessel, such as a clarifier. In that vessel, clean water will overflow for release, whereas settled metal precipitates (sludge) will be recycled to the acid mine drainage treatment tank, with a sludge-wasting side stream. A number of variations of this process exist, as dictated by the chemistry of ARD, its volume, and other factors. Generally, the products of the HDS process also contain gypsum and unreacted lime, which enhance both its settleability and resistance to re-acidification and metal mobilization.

Less complex variants of this process, such as simple lime neutralization, may involve no more than a lime silo, mixing tank and settling pond. These systems are far less costly to build, but are also less efficient (i.e., longer reaction times are required, and they produce a discharge with higher trace metal concentrations, if present). They would be suitable for relatively small flows or less complex acid mine drainage.

Calcium Silicate Neutralization

A calcium silicate feedstock, made from processed steel slag, can also be used to neutralize active acidity in AMD systems by removing free hydrogen ions from the bulk solution, thereby increasing pH. As the silicate anion captures H^+ ions (raising the pH), it forms monosilicic acid (H_4SiO_4), a neutral solute. Monosilicic acid remains in the bulk solution to play many roles in correcting the adverse effects of acidic conditions. In the bulk solution, the silicate anion is very active in neutralizing H^+ cations in the soil solution. While its mode-of-action is quite different from limestone, the ability of calcium silicate to neutralize acid solutions is equivalent to limestone as evidenced by its CCE value of 90-100% and its relative neutralizing value of 98%.

In the presence of heavy metals, calcium silicate reacts in a different manner than limestone. As limestone raises the pH of the bulk solution, and if heavy metals are present, precipitation of the metal hydroxides (with extremely low solubilities) is normally accelerated and the potential of armoring of limestone particles increases significantly. In the calcium silicate aggregate, as silicic acid species are absorbed onto the metal surface, the development of silica layers (mono- and bi-layers) lead to the formation of colloidal complexes with neutral or negative surface charges. These negatively charged colloids create an electrostatic repulsion with each other (as well as with the negatively charged calcium silicate granules) and the sequestered metal colloids are stabilized and remain in a dispersed state - effectively interrupting metal precipitation and reducing vulnerability of the material to armoring.

Carbonate Neutralization

Generally, limestone or other calcareous strata that could neutralize acid are lacking or deficient at sites that produce acidic rock drainage. Limestone chips may be introduced into sites to create a neutralizing effect. Where limestone has been used, such as at Cwm Rheidol in mid Wales, the positive impact has been much less than anticipated because of the creation of an insoluble calcium sulfate layer on the limestone chips, binding the material and preventing further neutralization.

Ion Exchange

Cation exchange processes have previously been investigated as a potential treatment for acid mine drainage. The principle is that an ion exchange resin can remove potentially toxic metals (cationic resins), or chlorides, sulfates and uranyl sulfate complexes (anionic resins) from mine water. Once the contaminants are adsorbed, the exchange sites on resins must be regenerated, which typically requires acidic and basic reagents and generates a brine that contains the pollutants in a concentrated form. A South African company that won the 2013 IChemE (ww.icheme.org) award for water management and supply (treating AMD) have developed a patented ion-exchange process that treats mine effluents (and AMD) economically.

Constructed Wetlands

Constructed wetlands systems have been proposed during the 1980s to treat acid mine drainage generated by the abandoned coal mines in Eastern Appalachia. Generally, the wetlands receive near-neutral water, after it has been neutralized by (typically) a limestone-based treatment process. Metal precipitation occurs from their oxidation at near-neutral pH, complexation with organic matter, precipitation as carbonates or sulfides. The latter results from sediment-borne anaerobic bacteria capable of reverting sulfate ions into sulfide ions. These sulfide ions can then bind with heavy metal ions, precipitating heavy metals out of solution and effectively reversing the entire process.

The attractiveness of a constructed wetlands solution lies in its relative low cost. They are limited by the metal loads they can deal with (either from high flows or metal concentrations), though current practitioners have succeeded in developing constructed wetlands that treat high volumes and/or highly acidic water (with adequate pre-treatment). Typically, the effluent from constructed wetland receiving near-neutral water will be well-buffered at between 6.5-7.0 and can readily be discharged. Some of metal precipitates retained in sediments are unstable when exposed to oxygen (e.g., copper sulfide or elemental selenium), and it is very important that the wetland sediments remain largely or permanently submerged.

An example of an effective constructed wetland is on the Afon Pelena in the River Afan valley above Port Talbot where highly ferruginous discharges from the Whitworth mine have been successfully treated.

Precipitation of Metal Sulfides

Most base metals in acidic solution precipitate in contact with free sulfide, e.g. from H_2S or NaHS. Solid-liquid separation after reaction would produce a base metal-free effluent that can be discharged or further treated to reduce sulfate, and a metal sulfide concentrate with possible economic value.

As an alternative, several researchers have investigated the precipitation of metals using biogenic sulfide. In this process, Sulfate-reducing bacteria oxidize organic matter using sulfate, instead of oxygen. Their metabolic products include bicarbonate, which can neutralize water acidity, and hydrogen sulfide, which forms highly insoluble precipitates with many toxic metals. Although promising, this process has been slow in being adopted for a variety of technical reasons.

Technologies

Many technologies exist for the treatment of AMD from traditional high cost water treatment plants to simple in situ water treatment reagent dosing methods.

Metagenomic Study of Acid Mine Drainage

With the advance of Large-scale sequencing strategies, genomes of microorganisms in the acid mine drainage community are directly sequenced from the environment. The nearly full genomic constructs allows new understanding of the community and able to reconstruct their metabolic pathways. Our knowledge of Acidophiles in acid mine drainage remains rudimentary: we know of many more species associated with ARD than we can establish roles and functions.

Microbes and Drug Discovery

Scientists have recently begun to explore acid mine drainage and mine reclamation sites for unique soil bacteria capable of producing new pharmaceutical leads. Soil microbes have long been a source for effective drugs and new research, such as that conducted at the Center for Pharmaceutical Research and Innovation, suggests these extreme environments to be an untapped source for new discovery.

Acid Mine Drainage

The main cause of acid mine drainage is oxidation of iron pyrite. Iron pyrite, FeS2, is the most widespread of all sulfide minerals and is found in many ore bodies. During mining operations, particularly coal mining, iron pyrite in the ore is exposed to air and water, causing it to be oxidized to sulfuric acid and ferrous ion:

$$FeS_2 + \tfrac{7}{2} O_2 + H_2O \leftrightarrow Fe^{2+} + 2SO_4^{2-} + 2H^+. \quad (1)$$

There is almost always enough moisture in mine wastes and mine workings to allow above reaction (1) to occur releasing acidity and dissolved ferrous ion into the water. Next, dissolved Fe^{2+} is oxidized slowly by dissolved oxygen to Fe3+, consuming some acidity:

$$Fe^{2+} + \tfrac{1}{4} O_2 + H^+ \leftrightarrow Fe^{3+} + \tfrac{1}{2} H_2O. \quad (2)$$

Above pH 4 and in the absence of iron-oxidizing bacteria, equn (2) is the rate-limiting step in the reaction sequence.

However, below pH 4 and in the presence of iron-oxidizing bacteria, the rate of equn (2) is greatly accelerated by a million-fold or more. Ferric ion formed in equn (2) can further oxidize pyrite, where ferric ion is reduced back to Fe^{2+}, releasing much more acidity.

$$FeS_2(s) + 14Fe^{3+} + 18H_2O \leftrightarrow 15Fe^{2+} + 2SO_4^{2-} + 16H^+. \quad (3)$$

By equn (3), eight times more acidity is generated when ferric ion oxidizes pyrite than when dissolved oxygen serves as the oxidant (16 equivalents of H^+ compared to 2 equivalents, per mole of FeS_2).

In the pH range from 2 to 7, pyrite oxidation by Fe^{3+} {equn (3)} is kinetically favored over abiotic oxidation by oxygen {equn(2)}.

In addition, equn (3) returns soluble Fe^{2+} to the reaction cycle.

Overall, 4 equivalents of acid are formed for each mole of FeS_2 oxidized in the cyclic reaction sequence of equn (2) and (3).

If bacterially mediated oxidation is occurring in equn (2), the reaction cycle can be accelerated by over a millionfold.

Ferric ion also hydrolyzes (reacts with water), releasing more acid to the water and forming insoluble ferric hydroxide, which can coat streambeds with the yellow-orange deposits known as yellow boy.

$$Fe^{3+} + 3H_2O \leftrightarrow Fe(OH)_3(s) + 3H^+.$$

$Fe(OH)_3$ precipitates serve as a reservoir for dissolved Fe^{3+}.

If the generation of Fe^{3+} by equn (3) is stopped because of lack of oxygen, then Fe^{3+} is supplied by dissolution of solid $Fe(OH)_3$ and is available to react via equn (3).

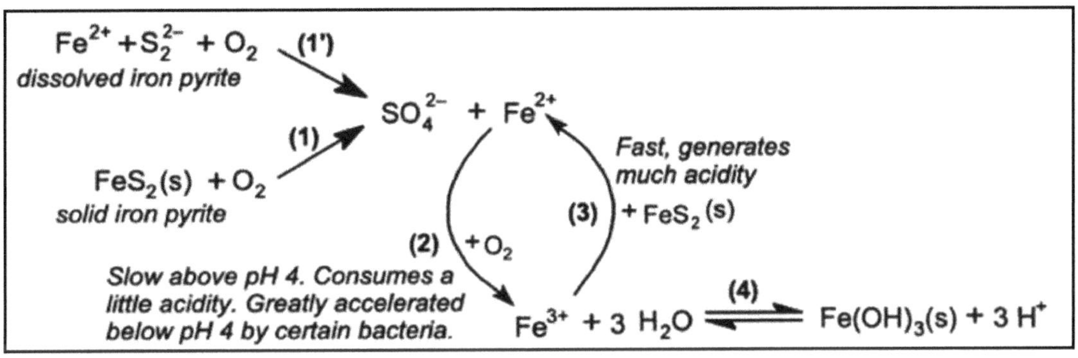

Step 1: Iron pyrite, dissolved or solid, is oxidized by dissolved oxygen equn (1), producing Fe^{3+}, SO_4^{2-}, and lowering the pH.

Step 2: Fe^{2+} formed in Step 1 is oxidized slowly by dissolved oxygen to Fe^{3+}, equn (2). This is the rate-limiting step in the reaction sequence in the absence of iron-oxidizing bacteria. The abiotic rate decreases with lower pH. However, iron-oxidizing bacteria can greatly accelerate this step when the pH falls below 4.

Step 3: Fe^{3+} from Step 2 is reduced rapidly back to Fe^{2+} by pyrite equn (3), generating much acidity. Ferrous ion, Fe^{2+} , generated in Step 3 re-enters the reactioncycle via step 2.

Step 4: A portion of ferric ion, Fe^{3+}, reacts with water to form $Fe(OH)_3$ precipitate, releasing more acidity equn (4).

When theFe^{3+} concentration diminishes, Fe(OH)3 precipitate can dissolve, acting as a reservoir for replenishing Fe^{3+} and maintaining the acid producing cycle.

As pH is lowered, Step 1 becomes less important and the abiotic rate of Step 2 decreases.

However, Step2 can be greatly accelerated by certain bacteria such as Metallogenium, Ferrobacillus,Thiobacillus, and Leptospirillum, which derive energy from the oxidation of Fe^{2+} to Fe^{3+}.

Below pH 4, these bacteria catalyze Step 2, speeding up the overall reaction rate by a factor as large as 106, and can lower the pH to 2 or less.

Furthermore, these bacteria can tolerate high concentrations of dissolved metals (e.g., 40,000 mg/L Zn and Fe; 15,000 mg/L Cu) before experiencing toxic effects.

They thrive in mine drainage waters as long as a minimal amount of oxygen is present. Once bacterial acceleration occurs, it is hard to reverse.

Non Metal Sulphides donot Generate Acidity

The oxidation by dissolved O_2 of noniron metal sulfides does not generate significant amounts of acidity. The metals are released as dissolved cations but acidity is not produced.

$$CuS(s)+2O_2 \leftrightarrow Cu^{2+}(aq)+SO_4^{2-}(aq).$$
$$ZnS(s)+2O_2 \leftrightarrow Zn^{2+}(aq)+SO_4^{2-}(aq).$$
$$PbS(s)+2O_2 \leftrightarrow Pb^{2+}(aq)+SO_4^{2-}(aq).$$

Two possible reasons for the lack of acid formation when noniron sulfides are oxidized are:

$$(iron\ pyrite)\quad S_2^{2-}+{}^7\!/_2\,O_2+H_2O \leftrightarrow 2SO_4^{2-}+2H^+$$
$$(other\ sulfides)\quad S^{2-}+2O_2 \leftrightarrow 2SO_4^{2-}$$

The oxidation state of sulfur is different in iron pyrite than in other sulfides, occurring as S_2^{2-} in iron pyrite and as S^{2-} in other sulfides. The respective oxidation reactions of these two sulfur forms indicate that acid is produced only with S_2^{2-}.

Cu^{2+}, Zn^{2+}, Pb^{2+}, etc., do not hydrolyze as extensively as does Fe^{3+}, so noniron sulfides do not react significantly by reactions.

$$Fe^{3+}+3H_2O \leftrightarrow Fe(OH)_3(s)+3H^+.$$

Ferric ion, Fe^{3+}, can oxidize other metal sulfides, such as ZnS (sphalerite), CuS (covellite), PbS (galena), and $CuFeS_2$ (chalcopyrite), in a similar fashion to its oxidation of FeS_2, releasing metal cations into the water without generating acidity.

Acid Base Potential of Soil

The acid-base potential (ABP) is a measure of how effectively the alkalinity (neutralization potential) in a solid sample can neutralize the acid-producing potential resulting from the presence of pyrite of the sample. The acid-base potential is equal to the equivalents of calcium carbonate ($CaCO_3$) in excess of the amount needed to neutralize the acid that could potentially be produced from oxidation of pyritic sulfur.

Example: Determining the Acid-Base Potential (ABP)

The ABP is calculated by

ABP = (alkalinity) – (31.25)(wt % pyritic sulfur)

Where, ABP is given in tons of $CaCO_3$-equivalents per 1000 tons of solid material. Any rock or earth material with an ABP of –5.0 represents a soil with a net potential deficiency of 5.0 tons $CaCO_3$/1000 tons material, and is defined as a potentially toxic material.

1. If the ABP is positive, leachate from the sample is likely to be basic.
2. If the ABP is negative, leachate is likely to be acidic.
3. If the ABP is –5, or more negative, the earth material may be defined as a potentially toxic material.

Pesticides and Fertilizers

Pesticides

- Pesticides are substances or mixture of substances intended for preventing, destroying, repelling or mitigating any pest.
- Green Revolution leads to the usage of pesticides, fertilizers herbicides not only on farms but in backyard gardens and on front lawns as well.

- Many of these chemicals were developed in the lab and are petroleum-based.
- These chemicals allowed farmers and gardeners of every stripe to exercise greater control over the plants they want to grow by enriching the immediate environment and warding off pests.
- At the same time as these synthetic chemicals are abs- orbed by soil and run-off into near by water ways causing land and water pollution.

Pesticides and Pollution of the Soil

- pesticides affect soil quality.
- Pesticides decrease biodiversity in the soil because they do not just kill the intended pest; they often kill many of the other small organisms present.
- When life in the soil is killed off, the soil quality deteriorates and this has a knock-on effect upon the retention of water.

- When pesticides kill off most of the active soil organisms, the complex interactions which result in good fertility break down.
- Plants depend on millions of bacteria and fungi to bring nutrients to their rootlets. When these cycles are disrupted plants become more dependent upon exact doses of chemical fertilizers at regular intervals.
- We get large but watery vegetables and fruits, which often lack taste and nutrients and may even contain pesticide residues.
- The misuse of pesticides can cause valuable pollinators such as bees and hover-flies to be killed and this in turn can badly affect food crops.
- There are fears that through natural selection, some pests may eventually become quite resistant to pesticides. Farmers may then try increasing amounts used of pesticides, making the problem worse.
- Pesticides are also polluting water and air.

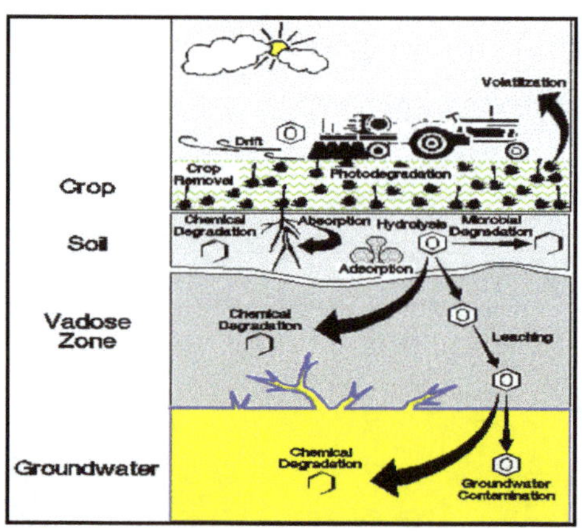

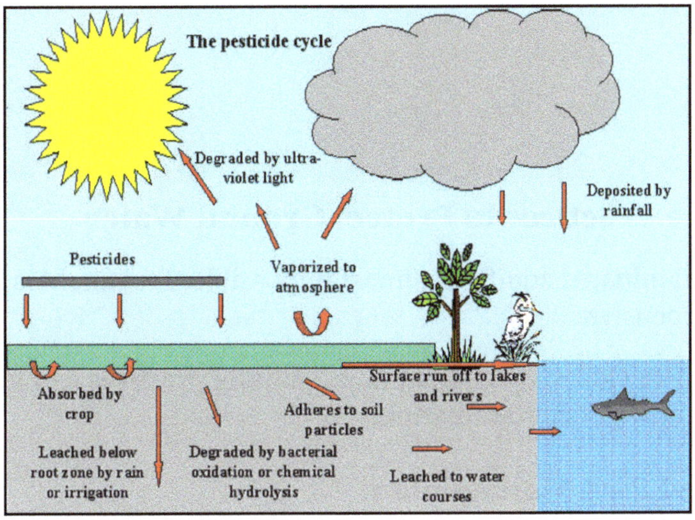

Pesticides and Pollution of the Water

- When the excess nutrients from all the fertilizer and pesticides we use runs off into our waterways, they cause algae blooms sometimes big enough to make waterways impassable.

- When the algae die, they sink to the bottom and decompose in a process that removes oxygen from the water. Fish and other aquatic species can't survive in these so-called "dead zones" and so they die or move on to greener underwater pastures.
- When those chemicals get into waterways, fish ingest them and become diseased. Humans who eat diseased fish can themselves become ill, completing the circle wrought by pollution.
- Pesticides also polluting the groundwater. As 90% of the population use ground water as drinking water it results in great health problems.
- Individual pesticides have unique properties, and many variable factors determine the specific risk in terms of water pollution.

Pesticide half-life

Mobility in soil

Solubility in water

Rainfall

Microbial activity

Soil temperature

Pesticide Handling Practices to Protect Ground Water

- Cleaning a contaminated aquifer is almost impossible. Treatment is complicated, time consuming, and expensive.
- The following pesticide management and handling practices can help you reduce the potential for groundwater contamination.

Practice Integrated Pest Management

Accurately identify the pest and apply pesticides only when necessary, using the lowest label rate that gives adequate pest control. Minimizing pesticide use makes good economic sense and reduces the potential for pest resistance and environmental problems.

Select Pesticides Carefully

Pesticides that are not adsorbed to soil particles can easily move in water, do not vaporize easily, are relatively stable, and thus have the greatest potential to leach into groundwater.

Consider the Vulnerability of the Area

Determine the relative susceptibility of the soil to leaching. Soil texture, organic matter content, and permeability affect pesticide movement. To the extent possible, determine the depth of the water table and the relative permeability of the geologic layers between the soil surface and the groundwater. Avoid spraying ditches and waterways along the field.

Consider the Location and Condition of Wells

Cap and seal wells properly and Slope the surrounding ground away from the wellhead to keep runoff away from the well.

Measure Accurately

Carefully calculate how much pesticide concentrate you need to treat the specific site with the equipment you plan to use. Handle pesticides carefully to avoid spills.

Prevent Back Siphoning

To prevent pesticides from siphoning back into the water supply, keep the end of the fill hose above the water level in the spray tank. Use an anti-backflow device (check valve) on the fill hose, especially when siphoning water directly from a pond or stream.

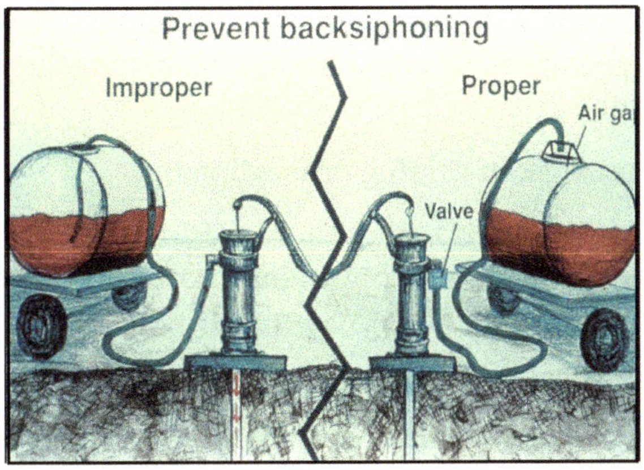

- Delay pesticide application if you expect rain because rainfall will increases pesticide run-off and leaching.
- Construct the pesticide storage area on a concrete floor sealed with an impervious material, located away from all water sources.
- The solution is to use organic materials like composted manure and other natural materials, as well as crop rotation, to help improve soil fertility, rather than synthetic fertilizers and pesticides that can result in soil and water pollution.

Soil & Water Pollution from Fertilizers

To Choose or Not To Choose

What are fertilizers?

- Fertilizer is any organic or inorganic material of natural or synthetic origin added to a soil to supply one or more plant nutrients essential to the growth of plants.
- Fertilizers come in various forms. The most typical form is solid fertilizer in granulated or powdered forms. The next most common form is liquid fertilizer; some advantages of liquid fertilizer are its immediate effect and wide coverage.
- Fertilizers are broadly divided into organic fertilizers (composed of organic plant or animal matter), inorganic or commercial fertilizers.

- Organic fertilizers include naturally occurring organic materials, (e.g. chicken litter, manure, worm castings, compost, seaweed, guano, bone meal) or naturally occurring mineral deposits (e.g. saltpeter).
- Fertilizers typically provide, in varying proportions:

  six macronutrients: nitrogen (N), phosphorus (P), potassium (K), calcium (Ca), magnesium (Mg), and sulfur (S); eight micronutrients: boron (B), chlorine (Cl), copper (Cu), iron (Fe), manganese (Mn), molybdenum (Mo), zinc (Zn) and nickel (Ni) (1987).
- Inorganic Fertilizers come in various form. Nitrogen fertilizers are often made using the Haber-Bosch process. Phosphate rock can also be processed into water-soluble phosphate (P_2O_5) with the addition of sulphuric acid (H_2SO_4) to make the phosphoric acid in phosphate fertilizers.

Pollution: Importance or Ignorance?

Is it the importance or our ignorance that we have overlooked such an important matter of fertilizers being a major cause of water and soil pollution.

Fertilizers have helped to increase crop production and keep pace with the increase in population but their growing use threatens our natural resources like water and soil.

BT Brinjal

Water Pollution

The nutrients, especially nitrates, in fertilizers can cause problems for natural habitats and for human health if they are washed off soil into watercourses or leached through soil into groundwater.

Eutrophication – Many definiton, one meaning.

- Eutrophication is defined as an increase in the rate of supply of organic matter in an ecosystem.
- "The term 'eutrophic' means well-nourished; thus, 'eutrophication' refers to natural or artificial addition of nutrients to bodies of water and to the effects of the added nutrients.... When the effects are undesirable, eutrophication may be considered a form of pollution.
- Eutrophication or more precisely hypertrophication, is the ecosystem response to the addition of artificial or natural substances, such as nitrates and phosphates, through fertilizers or sewage, to an aquatic system.

Fertilizers –> Eutrophicaton.

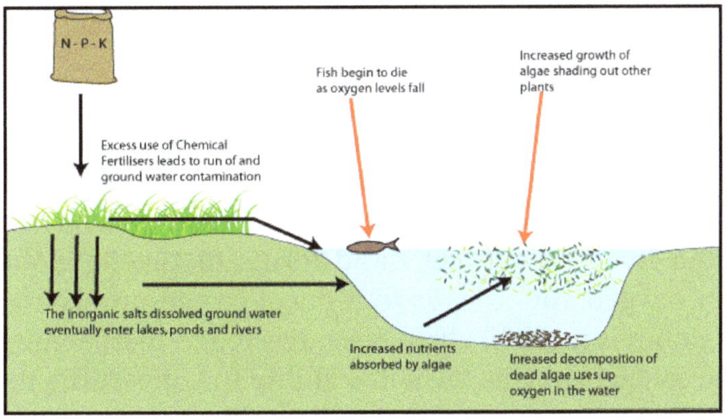

Negative environmental effects include hypoxia, the depletion of oxygen in the water, which induces reductions in specific fish and other animal populations. Other species (such as Nomura's jellyfish in Japanese waters) may experience an increase in population that negatively affects other species.

Impacts of Eutrophication

- In pure state water is odourless and tasteless .However, during the process of eutrophication algae or decaying aquatic vegetation can cause problems of odours.

- The world-wide occurrence of toxic algal blooms in fresh, marine and brackish waters creates problems for all life forms. For humans, the health risk of exposure to algal toxins comes above all from drinking water and from the recreational water use.
- Changes in the freshwater ecosystem caused by eutrophication indirectly affect the fish community. Low oxygen conditions can lead to the movement of fish out of affected areas. This leads to reduction of populations of these species, which are not able to tolerate new conditions.
- Livestock can also be at risk of cyanotoxins. There are documented many cases of lethal poisonings of livestock after drinking water with high content of toxic cyanobacterial bloom. Every year cattle deaths are associated with cyanobacterial blooms.

Drinking Water

- The main problem associated with high nitrate concentration in drinking water is the development of methaemoglobinaemia. The reactions can take place in stomach or in saliva. Bacteria present there can easily convert nitrate to nitrite. The nitrite combines with haemoglobin in red blood cells and forms methaemoglobin. Methaemoglobin is not able to carry oxygen and it leads to reduce oxygen uptake in the lungs. In pregnant women, haemoglobin is more likely to be converted to methaemoglobin. The people with deficiency of the special enzyme have low ability to convert the methaemoglobin back to haemoglobin.

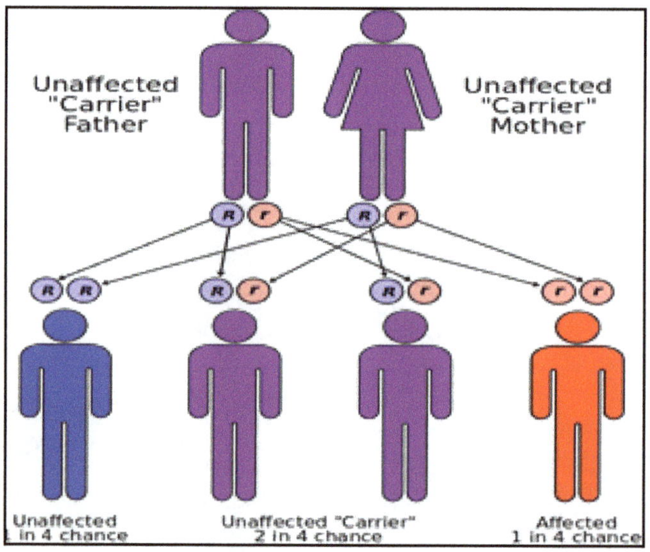

Congenital methemoglobinemia.

Blue Baby Syndrome

It is believed to be caused by high nitrate contamination in ground water resulting decreased oxygen carrying capacity of hemoglobin in babies leading to death. Infants under 6 months of age are particularly susceptible to methemoglobinemia caused by nitrates ingested in drinking water dehydration usually caused by gastroenteritis with diarrhea, sepsis, and topical anesthetics containing benzocaine or prilocaine.

- Nitrate and nitrite are able to react with some chemicals in the body and form compounds called N-nitrosamines. These compounds have been proved to cause cancer in animals, which suggests that they may also cause cancer in humans.

Soil Pollution

- Soil acidification is the build-up of hydrogen cations, also called protons, reducing the soil such as aluminium sulfate, which reacts in the soil to release protons. Many nitrogen compounds, which are added as fertilizer, acidify soil over the long term because they produce nitrous and nitric acid when oxidized in the process of nitrification.

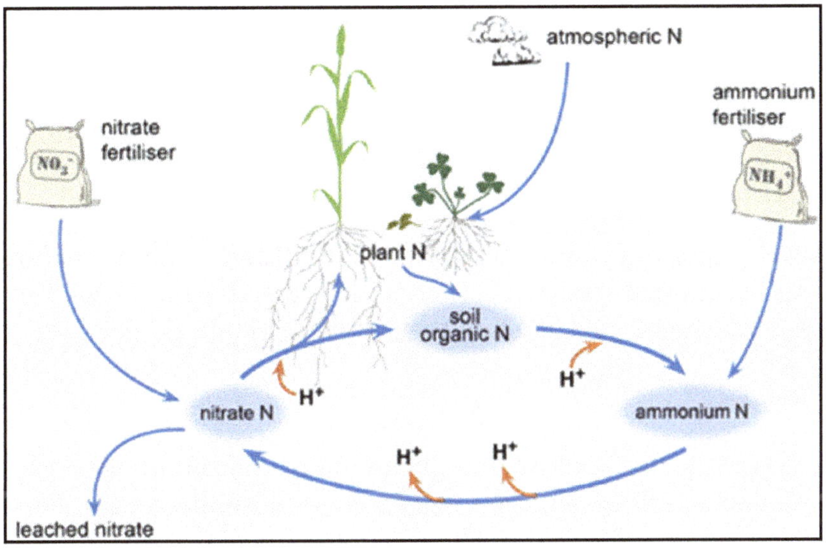

- Contamination with impurities such as fluorides, cadmium and uranium.

- The concentration of up to 100 mg/kg of cadmium in phosphate minerals (for example, minerals from Nauru and the Christmas islands increases the contamination of soil with cadmium, for example in New Zealand.

- Radioactive Polonium-210 contained in phosphate fertilizers is absorbed by the roots of plants and stored in its tissues; tobacco derived from plants fertilized by rock phosphates contains Polonium-210 which emits alpha radiation estimated to cause about 11,700 lung cancer deaths each year worldwide/Uranium In 2008, CSIC scientists said that nearly half of the catfish showed signs of alarming levels of mercury, and 15-20 per cent exceeded the maximum legal limits. The mercury contamination originates from the Flix swam. The mercury contamination originates from the Flix swamp, where CSIC detected some 700,000 tonnes of toxic mud in 2004.

- Toxic persistent organic pollutants ("POPs"), such as Dioxins, polychlorinated dibenzo-p-dioxins (PCDDs), and polychlorinated dibenzofurans (PCDFs) have been detected in agricultural fertilizers and soil amendments. Persistent organic pollutants (POP s) are organic compounds that are resistant to environmental degradation through chemical, biological, and photolytic processes. POP exposure can cause death and illnesses including disruption of the endocrine, reproductive, and immune systems.

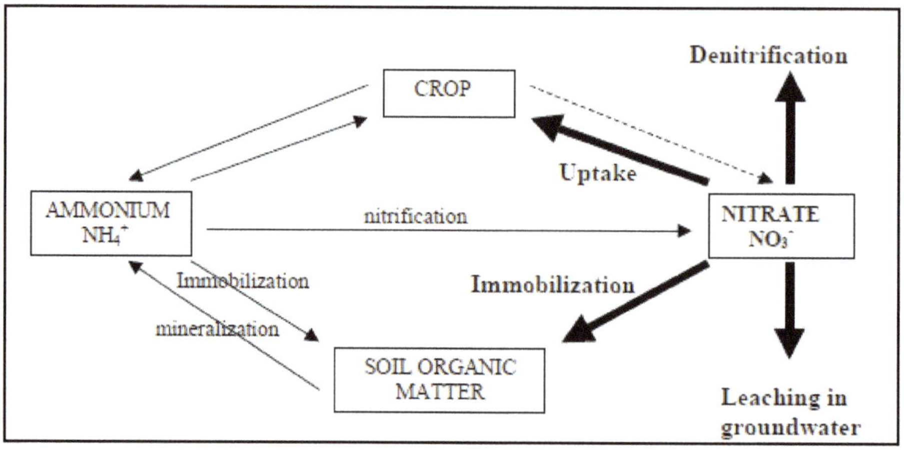

The Effect of Some Common Synthetic Fertilizers on the Soil

Ammonium Sulfate

In the soil, reacts with water to produce sulfuric acid. Sulfuric acid has a pH of less than 1 (2). It is extremely toxic and kills organisms. Hydrogen ions released from the acid replace alkaline elements on the cation exchange sites, depleting the soil of nutrients.

Ammonium Nitrate

In the soil, breaks down into ammonium and nitrate. The ammonium is consumed by plants and fungi, or by denitrifying bacteria which eventually convert it to nitrate. The nitrates are consumed by soil organisms, leached, or converted to nitrogen gas and volatized.

Urea Formaldehyde

The formaldehyde dissolves in water and is leached through the soil. It is a highly toxic substance, killing all soil organisms it comes in contact with. Formaldehyde is used to preserve laboratory specimens.

Urea

The urea is consumed by bacteria which convert it to (excrete) anhydrous ammonia (which is a gas) and carbon dioxide (= $2(NH_3) + CO_2$) (9). Anhydrous ammonia is highly toxic and kills organisms. If urea is incorporated into the soil, the ammonia gas reacts with water (H_2O) to produce ammonium hydroxide (NH_4OH), which has a pH of 11.6 (12). It is highly caustic and causes severe burns. This creates a toxic zone in the immediate vicinity of the applied urea that kills seeds, seedlings and soil dwelling organisms.

References

- Suthar, S; Bishnoi, P; Singh, S; et al. (2009). "Nitrate contamination in groundwater of some rural areas of Rajasthan, India". Journal of Hazardous Materials. 171 (1–3): 189–199. doi:10.1016/j.jhazmat.2009.05.111
- "Research | AMRF/ORV Alguita Research Projects" Algalita Marine Research Foundation. Macdonald Design. Retrieved 13, February 2020

- State Coalition for Remediation of Drycleaners (2011). "A Citizen's Guide to Drycleaner Cleanup" (PDF). Retrieved 23, June 2020
- Fawell, J; Bailey, K; Chilton, J; Dahi, E (2006). Fluoride in drinking-water (PDF). Geneva: IWA for WHO. ISBN 9241563192
- "Lack of data on fracking spills leaves researchers in the dark on water contamination". StateImpact Pennsylvania. Retrieved 17, April 2020
- ^ATSDR (US Agency for Toxic Substance & Disease Registry) (2008). "Follow-up Health Consultation: Anniston Army Depot." (PDF). Retrieved 23, January 2020
- World Health Organization (WHO) (2006). "Protecting Groundwater for Health - Understanding the drinking-water catchment" (PDF). Retrieved 28, May 2020
- Lin SD, Lee CC (2001). Water and Wastewater Calculations Manual. McGraw Hill Professional. pp. 13–. ISBN 978-0-07-137195-7
- "About the Organisation for Economic Co-operation and Development (OECD)" Organisation for Economic Co-operation and Development, Retrieved 09, August 2020
- "Texas fracking site that spilled 42,000 gallons of fluid into residential area hopes to reopen". RT International. Retrieved 2016-05-07
- Smith, M; Cross, K; Paden, M; Laben, P, eds. (2016). Spring - managing groundwater sustainably (PDF). IUCN. ISBN 978-2-8317-1789-0
- Xu, Y; Usher, B, eds. (2006). Groundwater pollution in Africa (PDF). Taylor & Francis. ISBN 0-415-41167-X
- Ross, N, ed. (2010). Clearing the waters a focus on water quality solutions. Nairobi, Kenya: UNEP. ISBN 978-92-807-3074-6
- United Nations Environment Programme (UNEP) (2015). "Good Practices for Regulating Wastewater Treatment" (PDF). Retrieved 17, January 2020
- Ahnert, A.; Borowski, C. (2000). "Environmental risk assessment of anthropogenic activity in the deep-sea". Journal of Aquatic Ecosystem Stress and Recovery. 7 (4): 299–315. doi:10.1023/A:1009963912171
- Pfunt,H; Houben,G; Himmelsbach,T (2016). "Numerical modeling of fracking fluid migration through fault zones and fractures in the North German Basin". Hydrogeology Journal. 24 (6): 1343–1358. doi:10.1007/s10040-016-1418-7

Permissions

All chapters in this book are published with permission under the Creative Commons Attribution Share Alike License or equivalent. Every chapter published in this book has been scrutinized by our experts. Their significance has been extensively debated. The topics covered herein carry significant information for a comprehensive understanding. They may even be implemented as practical applications or may be referred to as a beginning point for further studies.

We would like to thank the editorial team for lending their expertise to make the book truly unique. They have played a crucial role in the development of this book. Without their invaluable contributions this book wouldn't have been possible. They have made vital efforts to compile up to date information on the varied aspects of this subject to make this book a valuable addition to the collection of many professionals and students.

This book was conceptualized with the vision of imparting up-to-date and integrated information in this field. To ensure the same, a matchless editorial board was set up. Every individual on the board went through rigorous rounds of assessment to prove their worth. After which they invested a large part of their time researching and compiling the most relevant data for our readers.

The editorial board has been involved in producing this book since its inception. They have spent rigorous hours researching and exploring the diverse topics which have resulted in the successful publishing of this book. They have passed on their knowledge of decades through this book. To expedite this challenging task, the publisher supported the team at every step. A small team of assistant editors was also appointed to further simplify the editing procedure and attain best results for the readers.

Apart from the editorial board, the designing team has also invested a significant amount of their time in understanding the subject and creating the most relevant covers. They scrutinized every image to scout for the most suitable representation of the subject and create an appropriate cover for the book.

The publishing team has been an ardent support to the editorial, designing and production team. Their endless efforts to recruit the best for this project, has resulted in the accomplishment of this book. They are a veteran in the field of academics and their pool of knowledge is as vast as their experience in printing. Their expertise and guidance has proved useful at every step. Their uncompromising quality standards have made this book an exceptional effort. Their encouragement from time to time has been an inspiration for everyone.

The publisher and the editorial board hope that this book will prove to be a valuable piece of knowledge for students, practitioners and scholars across the globe.

Index

CPSIA information can be obtained
at www.ICGtesting.com
Printed in the USA
LVHW062014100122
708212LV00002B/222

9 781639 891740